教育部中等职业教育专业技能课立项教材

旅游服务与酒店管理类

西餐英语

（第二版）

主　编◎张艳红
副主编◎赵　静　郑　革
参　编◎（以编写章节为序）
钱　俊　姜　珊　刘欣雨　夏　孟
陈慧瑛　崔　朋　葛苗苗

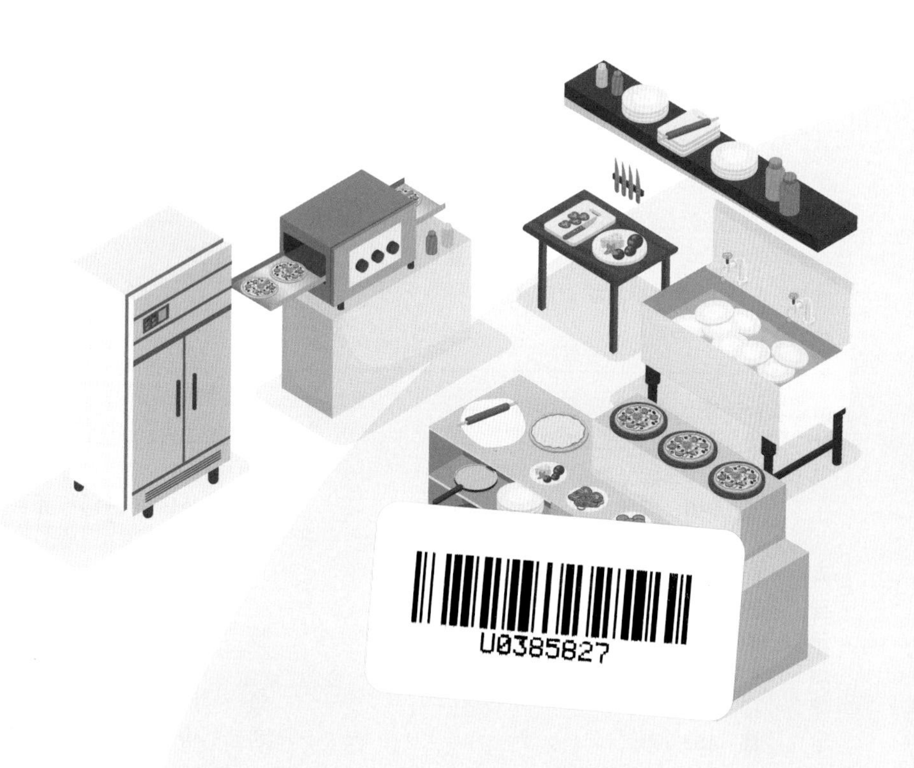

XICAN YINGYU

中国人民大学出版社
·北京·

图书在版编目（CIP）数据

西餐英语 / 张艳红主编 . -- 2 版 . -- 北京：中国人民大学出版社，2022.9
教育部中等职业教育专业技能课立项教材
ISBN 978-7-300-31012-1

Ⅰ. ①西… Ⅱ. ①张… Ⅲ. ①西式菜肴－英语－中等职业教育－教材 Ⅳ. ① TS972.118

中国版本图书馆 CIP 数据核字（2022）第 176667 号

教育部中等职业教育专业技能课立项教材
西餐英语（第二版）
主　编　张艳红
副主编　赵　静　郑　革
参　编（以编写章节为序）
钱　俊　姜　珊　刘欣雨　夏　孟
陈慧瑛　崔　朋　葛苗苗
Xican Yingyu

出版发行　中国人民大学出版社
社　　址　北京中关村大街 31 号　　邮政编码　100080
电　　话　010－62511242（总编室）　010－62511770（质管部）
　　　　　010－82501766（邮购部）　010－62514148（门市部）
　　　　　010－62515195（发行公司）　010－62515275（盗版举报）
网　　址　http://www.crup.com.cn
经　　销　新华书店
印　　刷　北京瑞禾彩色印刷有限公司
规　　格　185 mm × 260 mm　16 开本
印　　张　9.5
字　　数　165 000
版　　次　2012 年 5 月第 1 版
　　　　　2022 年 9 月第 2 版
印　　次　2022 年 9 月第 1 次印刷
定　　价　36.00 元

前言

PREFACE

《西餐英语》是中等职业学校烹饪专业英语课程用书，也可以作为旅行社的培训、烹饪从业人员和西餐美食爱好者的自学教材。本教材遵循以提升学生的职业能力为本位，将行业知识与职业技能渗透在英语教学中，旨在培养学生掌握必要的专业词汇，熟悉烹饪从业人员在厨师岗位中交流的基本技巧，并通过课后练习等环节使学生可以检验自己的学习效果。

本教材共分 9 个单元，每个单元编写体例一致，均包含 Learning goals，Word list，Lead in，Focus on learning，Get if off，Unit exercises，Learning tips，Culture knowledge 八个部分，设计意图如下：

1. Learning goals，明确本单元知识重点和能力要点以及学习后应达到的水平，对学生了解预期学习结果具有明确的导向和激励作用，是教学活动的出发点与归宿，是对学生进行学习评价的主要内容和标准之一。

2. Word list，列出了本单元使用的词汇，并且配以相关的图片，便于学生学习和直观认识。

3. Lead in，以图片连线等方式引入新课，将学生带入现实的工作环境中，模拟真实的工作场景，激发学生学习专业英语的热情。

4. Focus on learning，采用情景模拟、角色扮演、任务驱动等多种方法学习单词、句型和对话，呈现形式以听、读、说为主。引导学生在掌握单词的基础上模拟真实的工作场景和内容进行对话练习，同时引导学生进行阶段性自我学习情况评价，增强学生学习专业英语的信心。

5. Give if off，重点词组和句型拓展练习，提高学生的语言运用能力。

6. Unit exercises，巩固本单元所学的知识点。

7. Learning tips，以“小贴示”的形式拓展烹饪专业知识，提高学生的学习兴趣与实践能力。

8. Culture knowledge，了解西餐美食背后的文化知识，在学习掌握专业英语技能的基础上开拓文化视野。

建议每周安排 2 个学时，预计安排 36 周，共 72 学时。

本教材由北京市朝阳区教育科学研究院英语教研员张艳红担任主编，北京市劲松职业高中赵静、郑革担任副主编。主要参编人员及编写章节如下：钱俊 Unit1、姜珊 Unit2、刘欣雨 Unit3、赵静 Unit4、夏孟 Unit5、陈慧瑛 Unit6、崔朋 Unit7、葛苗苗 Unit8 和张艳红 Unit9。

本教材在编写过程中，广泛吸取了国内外现有研究成果，引用了其中有关文献资料，同时得到了北京市劲松职业高中、北京市求实职业学校和北京市丰台区职业教育中心学校的大力支持；得到了业内人士林鹏涛和郑宏斌先生的鼎力协助和北京市劲松职业高中西餐烹饪专业主任郑革和王跃辉老师的大力帮助；本书的出版也融入了中国人民大学出版社编辑的心血，在此一并表示衷心的感谢！

由于编者水平有限，教材中难免有不妥、疏漏和值得商榷的地方，还望广大读者批评指正。

编者

2022 年 6 月

目录 CONTENTS

Unit 1

Title Used in the Kitchen & Floor Plan

Learning goals（学习目标）

You will be able to

★ know some job titles and floor plan in the kitchen.

★ know the uniforms in the kitchen.

★ describe one's job.

Word list（单词表）

Key words & expressions（重点单词和词组）

job titles ['taɪtl]	职位	floor plan	楼层平面图
uniform ['juːnɪfɔːm]	制服	describe [dɪ'skraɪb]	描述
trainee [ˌtreɪ'niː]	学员	manager ['mænɪdʒə (r)]	经理
kitchen ['kɪtʃɪn] helper	帮厨	cook	厨师
executive [ɪg'zekjətɪv] chef	行政总厨师长	executive sous ['suː]-chef	行政副总厨师长
Italian kitchen sous-chef	意大利厨房副厨师长	butchery ['bʊtʃəri] sous-chef	肉食加工副厨师长
pastry ['peɪstri] & bakery ['beɪkəri] sous-chef	西点厨房副厨师长	commissary ['kɒmɪsəri] sous-chef	备餐厨房副厨师长
banquet ['bæŋkwɪt] kitchen sous-chef	宴会厨房副厨师长	chief [tʃiːf] steward '[stjuːəd]	管事部经理
be in charge [tʃɑːdʒ] of	对……负责	main [meɪn] kitchen	主厨房
chef [ʃef] hat	厨师帽	tie	三角巾
towl ['tauəl]	手巾	apron ['eɪprən]	围裙
jacket	上衣	black or plaid [plæd] pants	厨师黑格裤
shoes	鞋	commis chef	厨师
demi chef	领班	chef de partie ['pɑːtɪ]	厨师主管
steward ['stjuːəd]	厨房清洁工	butcher ['bʊtʃə (r)]	屠夫

Expanded words & expressions（拓展单词和词组）

What do you think of...?	你认为……怎么样？	attractive [ə'træktɪv]	有吸引力的
boring	令人厌烦的，无聊的	satisfying ['sætɪsfaɪŋ]	令人满意的

Lead in（导入）

Activity 1 Self-introduction.（自我介绍）

Hello, I am Li Ming. My English name is Mike. I'm from Guangming Vocational School. I'll go to the Grand Hotel as a trainee for a year. My chef is Mr. Wang. He is very kind and strict. Now follow me to the kitchen, please.

Activity 2 Look and match.（看图并连线）

kitchen helper | cook | manager | trainee

Activity 3 Oral practice.（口语练习）

A: What's your job?

B: I am a ____________________.

kitchen helper

cook

trainee

manager

Focus on learning（聚焦学习）

Part One

Activity 1 Read.（读词组）

executive chef 行政总厨师长	executive sous-chef 行政副总厨师长
coffee shop kitchen sous-chef 咖啡厅副厨师长	commissary sous-chef 备餐厨房副厨师长
banquet kitchen sous-chef 宴会厨房副厨师长	pastry & bakery sous-chef 西点副厨师长
Italian kitchen sous-chef 意大利厨房副厨师长	chief steward 管事部经理

Activity 2 Look and write.（看图写中文）

Notes:

总厨师长一般下设七个副厨师长，请和我一起看着这棵大树来学习吧。

executive chef ________

executive sous-chef ________

coffee shop kitchen sous-chef ________

commissary sous-chef ________

banquet kitchen sous-chef ________

pastry & bakery sous-chef ________

Italian kitchen sous-chef ________

chief steward ______

……

Activity 3 Look and write.（看图写英文）

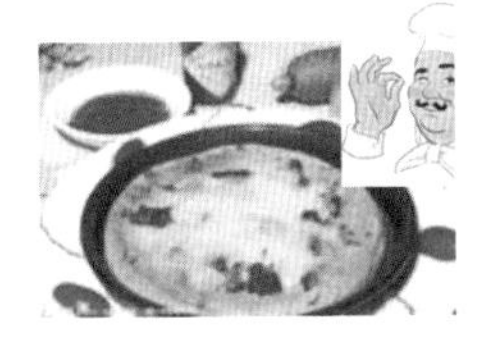

Activity 4 Read and answer.（读对话，回答问题）

Questions: 1. What does he do in the kitchen?
2. What is he in charge of?

Mike : What does he do in the kitchen?
Chef : He is a pastry & bakery sous-chef.
Mike : What is he in charge of?
Chef : He is in charge of making pastries.

A: Mike, what does he do in the kitchen?
B: He is a __________.
A: What is he in charge of?
B: He is in charge of ________________.
（making pizza, making coffee）

Activity 5 Listen and complete.（听录音，完成对话）

A. What do you think of your job	B. How many hours do you work every day
C. I find it attractive	D. How long do you work in the kitchen

Mike : ______________________________?
Chef : About 20 years.
Mike : ______________________________?
Chef : Nearly 9 hours.
Mike : ______________________________?
Chef : ______________________(boring, hard, interesting).

Activity 6 Pair work.（对话练习）

Mike is talking about________ (the titles) with Tom.

Mike

Tom

Part Two

Activity 1 Read.（读词组）

main kitchen 主厨房
hot kitchen 热菜厨房
commissary kitchen 备餐厨房
Chinese kitchen 中餐厨房
Coffee shop kitchen 咖啡厅厨房
cold kitchen 冷菜厨房
butchery kitchen 肉食厨房
pastry & bakery kitchen 西点厨房
Italian kitchen 意大利厨房

Activity 2 Write.（写出不同厨房的中英文表达）

main kitchen ________

Notes:

主厨房一般包括冷菜厨房、热菜厨房、肉食厨房、备餐厨房、西点厨房等，主厨房中准备的菜品会源源不断地供给以下不同的厨房。

cold kitchen	hot kitchen	butchery kitchen	commis-sary kitchen	pastry & bakery kitchen	……
________	________	________	________	________	

________ 主厨房

- ________ 咖啡厅厨房
- ________ 意大利厨房
- ________ 中餐厨房
- ……

Activity 3 Complete and read.（完成并朗读对话）

A. They are working for the Italian kitchen

B. He is in the commissary kitchen

C. Where is the commissary sous-chef?

D. He is in the coffee shop kitchen

Mike :Where is the coffee shop kitchen sous-chef?
Chef : ______________________.
Mike : ______________________?
Chef : ______________________.
Mike : What are they doing?
Chef : ______________________.

Activity 4 Decide true or false.（判断正误）

(　　) 1. The executive chef is in the main kitchen.

(　　) 2. The commissary sous-chef is in the butchery.

(　　) 3. The butchery sous-chef is in the butchery kitchen.

(　　) 4. The Italian kitchen sous-chef is in the commissary kitchen.

(　　) 5. The coffee shop kitchen sous-chef is in the coffee garden.

(　　) 6. The pastry & bakery sous-chef is in the pastry & bakery kitchen.

Part Three

Activity 1 Read.（读单词和词组）

chef hat 厨师帽　　tie 三角巾　　towel 手巾　　apron 围裙

jacket 上衣　　black or plaid pants 黑裤或格子裤　　shoes 鞋

Activity 2 Look and write.（看图写英文）

Notes:

每家酒店的厨师上衣都是不一样的，一般会各自设计统一的款式。厨师长通常穿黑色的厨师裤子，而普通厨师通常穿格子的厨师裤子。

Activity 3 Fill in and evaluate.（填空并自我评价）

1. sous-ch_ f　　2. ex_cutive　　3. jack_t　　4. ch_rge　　5. c_ffee

6. b_nquet　　7. p_ stry　　8. apr_n　　9. commiss_ry　　10. kitch_n

Assessment : If you can write 8-10 words, you are perfect.

If you can write 4-7 words, you are good.

If you can write 1-3 words, you will try again.

Give it off（知识拓展）

Practice 1 Match.（连线）

Practice 2 Summarize.（总结本单元的重点句型）

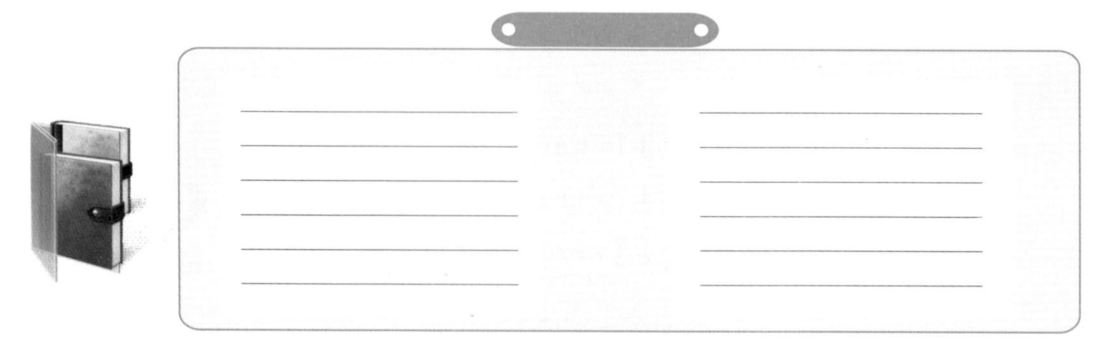

Practice 3 Write.（按照职别大小写出咖啡厅的职位）

coffee shop kitchen sous-chef 咖啡厅副厨师长	trainee 学员
commis chef 厨师	kitchen helper 帮厨
demi chef 领班	chef de partie 厨师主管

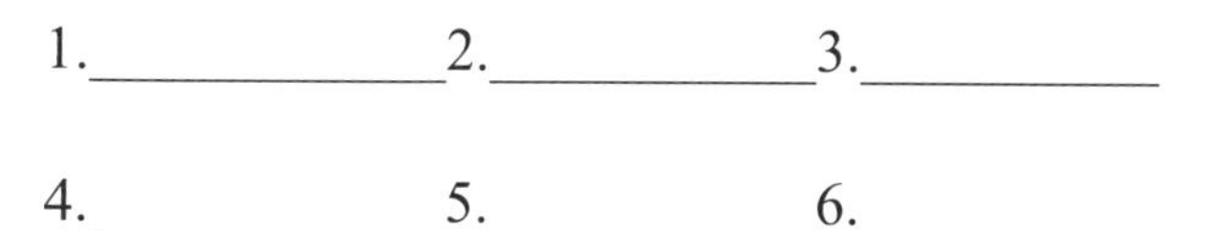

1.______________ 2.______________ 3.______________

4.______________ 5.______________ 6.______________

Notes:

一般而言，副厨师长之下设五个职别。这里以 coffee shop kitchen sous-chef 为例进行部门职别排序的练习。

Practice 4 Write and read.（写出工作名称并进行对话练习）

steward 厨房清洁工	butcher 屠夫
chef de partie 厨师主管	kitchen helper 帮厨

A : What do you do in the kitchen?

B: I'm a *steward*.

______________ ______________ ______________ ______________

Unit exercises（单元练习）

Exercise 1 Make dialogues.（对话练习）

A: What do you do in the kitchen? B: I'm a *cook*.

A: What are you doing? B: I am *making sauce*.

make sauce

make soup

grill sausage

make toast

cook fish

cook meat

make cake

cook chicken

make vegetables

Exercise 2 Tick and tell why.（选出你喜欢的工作并说出理由）

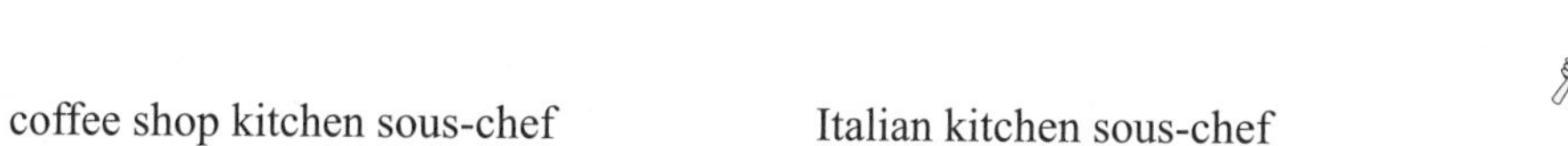

coffee shop kitchen sous-chef

Italian kitchen sous-chef

butchery sous-chef

pastry & bakery sous-chef

commissary sous-chef

banquet kitchen sous-chef

chief steward

executive sous-chef

butcher

trainee

demi chef

commis chef

Exercise 3 Make sentences in right order.（连词成句）

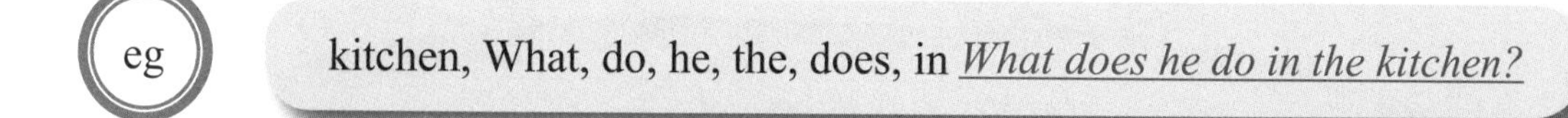

eg kitchen, What, do, he, the, does, in *What does he do in the kitchen?*

1. he, pastry & bakery kitchen,in charge of, is,

____.

2. is, the coffee garden, the coffee shop kitchen sous-chef, in

____.

3. making, the commis chef, sauce, is

____.

4. black, wears, the executive chef, pants

____.

5. this, find, job, attractive, I, is

____.

Exercise 4 Write in pairs.（写出你未来的理想工作）

I want to be *a coffee shop kitchen sous-chef.*

1. I want to be a /an ______________________________ .
2. We want to be ____________________________ .
3. You want to be a/an ______________________________ .
4. He wants to be a/an ______________________________.
5. They want to be ___________________________ .
6. She wants to be a/an ______________________________.

Learning tips（学习提示）

Kitchen Rules

☆ Wear uniform, wash hands, care personal hygiene.

☆ Can not bring food or beverage into the working section.

☆ Can not smoke indoors.

☆ Can not lend kitchen stuffs to others.

☆ Clean kitchen facilities and utensils regularly.

Culture knowleadge（文化知识）

西餐厨师岗位职责与素质要求

直接上级：西餐厨师领班

直接下属：无

岗位职责：

1. 按照菜品准确投料和烹制方法，制作西餐菜品。
2. 清理自己的工作台面，保持工作区域的清洁卫生，减少浪费。
3. 按照操作规格使用各种设备，清洁各种用具，并按照规定摆放好。

素质要求：

基本素质：有强烈的工作责任心。

自然条件：身体健康，品貌端正，年龄在 20 岁以上。

文化程度：厨师烹饪专业毕业。

外语水平：初级英语水平，具备与厨师英文沟通能力。

工作经验：有两年以上西餐制作经验。

特殊要求：具有中级技术等级证书。

Unit 2

Kitchen Facilities

Learning goals（学习目标）

You will be able to

★ get familiar with the facilities in the kitchen.

★ tell the names of the kitchen facilities.

★ describe the usages of the kitchen facilities.

Word list（单词表）

Key words & expressions（重点单词和词组）

oven ['ʌvn]	烤箱	gas stove [gæs stəʊv]	炉灶
grill [grɪl]	扒炉	microwave ['maɪkrəweɪv]	微波炉
salamander ['sæləmændə (r)]	明火焗炉	steamer ['sti:mə (r)]	蒸汽炉
dish washer	洗碗机	deep-fryer [di:p 'fraɪə (r)]	炸炉
slicer ['slaɪsə]	切片机	waffle iron ['wɒfl 'aɪən]	华夫炉
mincer ['mɪnsə (r)]	绞肉机	bone saw [bəʊn sɔ:]	切骨机
liquidizer ['lɪkwɪdaɪzə (r)]	榨汁机	stir machine [stɜ:(r) mə'ʃi:n]	搅拌机
tenderizer ['tendəraɪzə (r)]	软化机	kneading ['ni:dɪŋ] machine	揉面机
ice maker	制冰机	refrigerator [rɪ'frɪdʒəreɪtə (r)]	冰箱
ice cream machine	冰激凌机	freezer ['fri:zə (r)]	冰柜

Expanded words & expressions（拓展单词和词组）

fry [fraɪ]	煎，炒	grill [grɪl]	扒
liquidize ['lɪkwɪdaɪz]	榨汁	freeze [friːz]	冰冻
slice [slaɪs]	切片	tenderize ['tendəraɪz]	软化肉
mince [mɪns]	绞肉	tough [tʌf]	硬的
be used for...	用于做……	What about...?	……怎么样？

Lead in（导入）

Activity 1 Look and match.（看图并连线）

oven　　gas stove　　grill　　microwave

Activity 2 Decide true or false.（判断正误）

(　　) 1. Gas stove is a machine to cook food.

(　　) 2. Grill is a machine to grill meat.

(　　) 3. Microwave is a machine to cook food very quickly.

(　　) 4. Oven is a machine to heat or bake food.

Activity 3 Oral practice.（口语练习）

A: What is it?

B: It is a (an) ________________.

Focus on learning（聚焦学习）

Part One

Activity 1 Read.（读单词和词组）

salamander 明火焗炉　　steamer 蒸汽炉　　dish washer 洗碗机　　deep-fryer 炸炉

Activity 2 Look and write.（看图写英文）

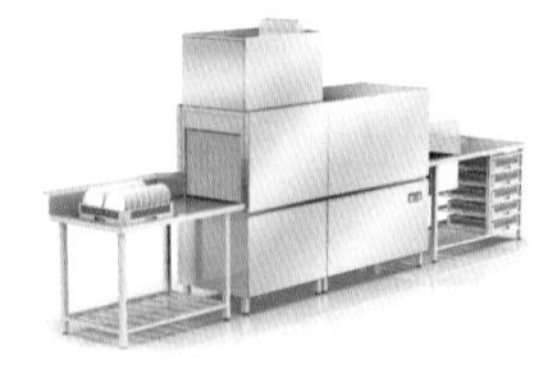

________　________　________　________

Activity 3 Read and answer.（读对话，回答问题）

Questions: 1. What do you call this machine in English?

2.What is it used for?

Mike : What do you call this in English?

Chef : It's a deep-fryer.

Mike : What is it used for?

Chef : It is used for frying food.

Mike : It is useful in the kitchen.

Chef : Yes, you are right.

A: What do you call this in English?

B: It is a (an) _________.

A: What is it used for?

B: It is used for _____________.

Activity 4 Listen and complete.（听录音，完成对话）

A .We use it to heat food	B.It's a steamer
C. making soup for a long time	D.You can fry it in the deep-fryer

Mike : What shall I call it?

Chef : ______________.

Mike : What is this used for?

Chef : ____________and it is good for ________.

Mike : Can I fry meat in it?

Chef : No, you can't. ___________.

Mike : I see.

Activity 5 Pair work.（对话练习）

eg Mike is talking about the dish washer with Tom.

Mike

Tom

Part Two

Activity 1 Read.（读单词和词组）

slicer 切片机	waffle iron 华夫饼模
mincer 绞肉机	bone saw 切骨机
liquidizer 榨汁机	stir machine 搅拌机
tenderizer 软化机	kneading machine 揉面机

Activity 2 Look, write and talk in pairs.（看图写英文并进行对话的练习）

A: What shall we call it?

B: It is a (an) ________.

 ________

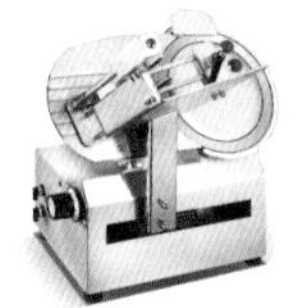 ________

 ________

 ________

 ________

 ________

 ________

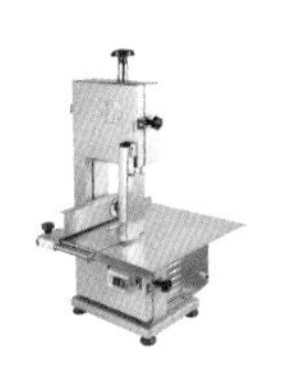 ________

Activity 3 Complete and read.（完成并朗读对话）

A. We can also liquidize them	B. you can liquidize vegetables, too
C. Put them in the liquidizer	D. Can you show me how to use it

Mike : How shall we do with these bananas?

Chef : ________________________.

Mike : OK. And what about the apples?

Chef : ________________________.

Mike : Only can it liquidize fruit?

Chef : No, ________________________.

Mike : Really? ________________________?

Chef : No problem.

Part Three

Activity 1 Tick.（勾出你在厨房中经常用到的冷冻设备）

ice maker 制冰机
refrigerator 冰箱
ice cream machine 冰激凌机
freezer 冷柜

Activity 2 Look and write.（看图并写出英文）

______ ______ ______ ______

Activity 3 Decide true or false.（判断正误）

1. ________Tenderizer is used for tenderizing tough meat.
2. ________We use the slicer to mince meat.
3. ________Oven is used for baking bread.
4. ________Refrigerator is a machine to keep food.

Activity 4 Fill in and evaluate.(填空并自我评价)

1. _lice	2. tenderiz_ _	3. min_e	4. st_ _mer	5. st_v_
6. gr_ll	7. m_ncer	8. f_y	9. ma_e	10. liqu_dizer

Assessment : If you can write 8-10 words, you are perfect.
If you can write 4-7 words, you are good.
If you can write 1-3 words, you will try again.

Give it off（知识拓展）

Practice 1 Match.（连线）

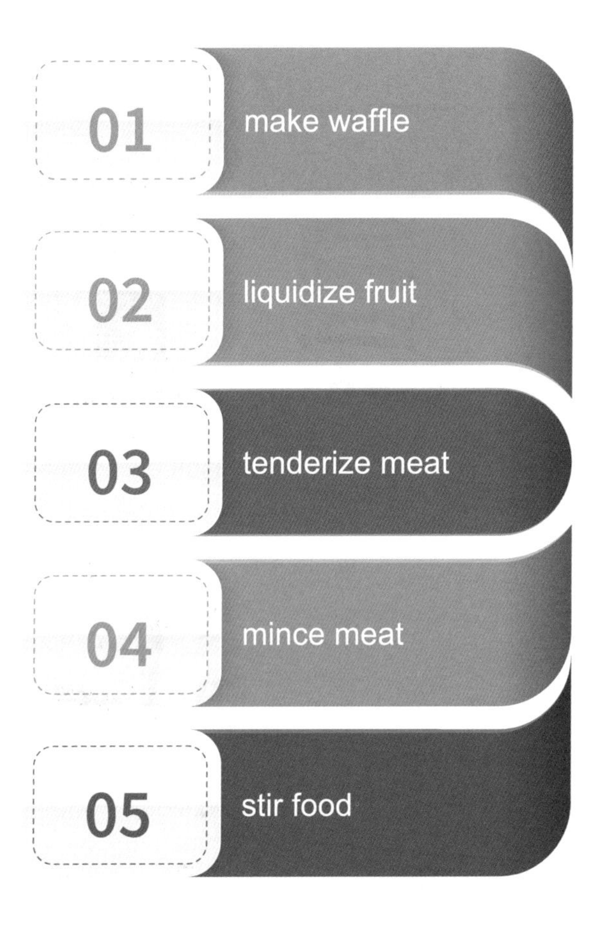

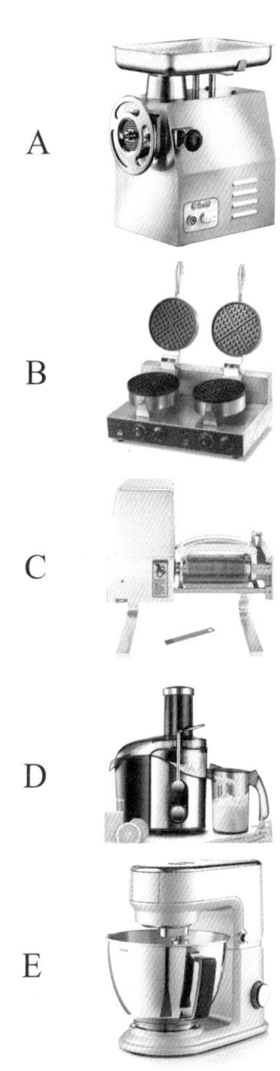

Practice 2 Summarize.（总结本单元的重点句型）

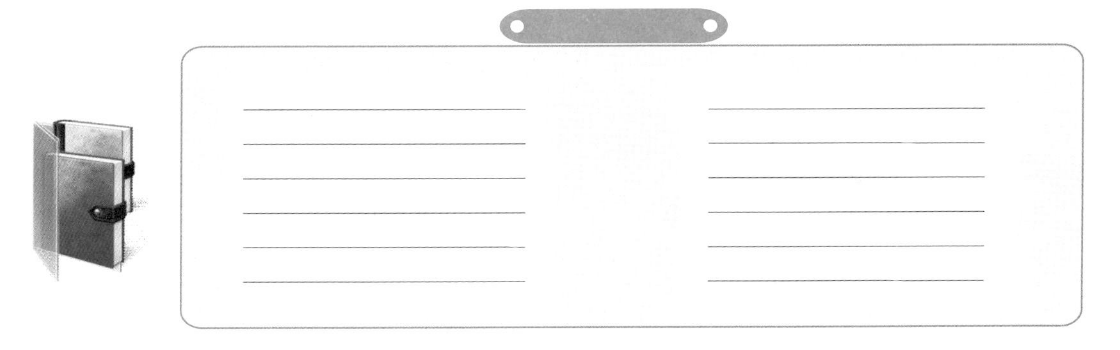

Practice 3 Make dialogues.（对话练习）

A: What do you call this in English?

B: It is a *refrigerator*.

A: What is it used for?

B: It is used for *keeping food*.

This is a *deep-fryer*.

It is used for *frying food*.

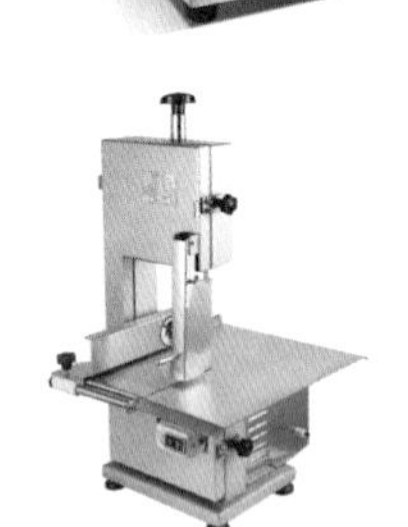

Unit exercises（单元练习）

Exercise 1 Choose the best answers.（选择正确答案）

1. I'd like to use the ____ to make a glass of orange juice.

   A. tenderizer　　B. mincer　　C. liquidizer　　D. slicer

2. We usually use the ______ to slice meat.

   A. tenderizer　　B. mincer　　C. liquidizer　　D. slicer

3. _______ is used to keep food at home.

   A. Freezer　　B. Ice maker

   C. Ice cream maker　　D. Refrigerator

4. I like eating ice cream. My dream is to have a (an)________ to make ice cream in my home.

   A. freezer　　B. ice maker

   C. ice cream machine　　D. refrigerator

5. There're many facilities in the kitchen. They're very ______.

   A. terrible　　B. difficult　　C. useful　　D. interesting

Exercise 2 Make sentences in right order.（连词成句）

1. in, useful, gas stove, the, is, the, kitchen

   ____.

2. ice maker, a, is, to, ice, make, machine

   ____.

3. tenderizer, is, used, for, tough, meat, tenderizing

   ____.

4. we, microwave, can, heat, food, use, to

   ____.

Exercise 3 Classify the facilities.（设备分类）

refrigerator	gas stove	deep-fryer	grill	slicer
steamer	waffle iron	mincer	freezer	oven
liquidizer	stir machine	tenderizer	bone saw	salamander
microwave	ice maker	ice cream machine	kneading machine	

炉灶设备	机械设备	制冷设备

Exercise 4 Translate into Chinese or English.（中英文翻译）

1. 榨汁机可以用来榨蔬菜和水果。

____

2. 我们可以用软化机来软化牛肉。

____

3. What do you call this in English? It's a salamander.

____

4. What shall we do with the deep-fryer?

____

5. What is the stir machine used for?

____

Learning tips（学习提示）

Kitchen safety is important, so tick the activities we should follow.

(　　　　)　　(　　　　)　　(　　　　)　　(　　　　)

A. Use hand closely to the machine.

B. Cut off the electricity while cleaning the electric socket.

C. Heat the food in the kitchen without people.

D. Wait for the maintainer to repair the electricity.

Culture knowledge（文化知识）

中餐和西餐的文化背景不同，烹饪方法和口味也有不同。对于中西餐厨房设备的需求也有不同。中餐厨房和西餐厨房设备相比肯定有区别，最明显的区别在于锅具，中餐厨房的锅具多是圆弧底的中式炒锅，而西餐厨房的锅具是平底锅，另外，中餐厨房还有一些只有中餐才会用到的厨具，比如砂锅、瓦罐等。一般的中餐厨房需要有蒸制米饭、馒头等主食时使用的蒸锅蒸柜；热火炒制菜品时使用的炉具灶具；小火煨煮时使用的传统砂锅器皿；腌制食品时使用的各式坛子等。

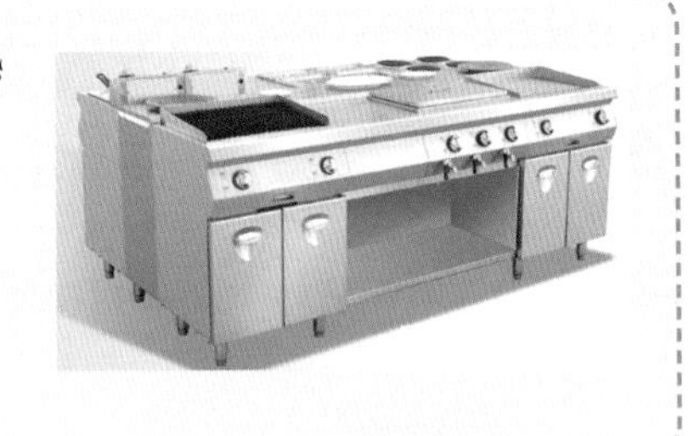

西餐在烹饪方式上相对简单一些，比较注重成熟度、调味方面的掌握。西餐烹饪方式主要有煎、炸、炒、煮、焖、烩、烤、焗、扒、铁板、串烧10余种，在厨房设备上使用更多的是组合式的西餐机械设备。一般来说包括炉具、扒炉、烤箱、电磁炉等。

Unit 3

Tools & Utensils

Learning goals（学习目标）

You will be able to

★ get familiar with the cooking utensils in the kitchen.

★ ask about the usage of cooking utensils.

★ describe the usage of cooking utensils.

Word list（单词表）

Key words & expressions（重点单词和词组）

braising pan [breɪz pæn]	焖锅	stockpot ['stɒkpɒt]	汤锅
frying pan ['fraɪɪŋ ˌpæn]	煎锅	grill pan [grɪl pæn]	烤肉盘
sauce pan [sɔs pæn]	沙司锅	sauté pan [sɒ'teɪ pæn]	嫩炒锅
crepe pan [kreɪp pæn]	煎薄饼锅	sugar pan ['ʃʊgə (r) pæn]	糖锅
whisk [wisk]	打蛋器	spider ['spaɪdər]	漏勺
mallet ['mælət]	肉锤	roastingfork ['rəʊstɪŋfɔ:(r) k]	烤肉叉
roasting tray ['rəʊstɪŋ treɪ]	烤盘	soup ladle ['leɪdl]	汤匙
serving spoon ['sɜrvɪŋ spu:n]	上菜匙	rolling-pin ['rəʊlɪŋ pɪn]	擀面杖
grater ['greɪtə (r)]	擦子	frying basket ['fraɪɪŋ 'bæskət]	油炸篮
colander ['kʌləndər]	沥水篮	potrack [pɒt ræk]	锅架
cutting board ['kʌtiŋ ˌbɔ:(r) d]	切菜板	cover ['kʌvə (r)]	餐盘罩
fish poacher [fɪʃ 'poʊtʃər]	蒸鱼盒	mixing bowl ['mɪksɪŋ ˌboʊl]	拌菜碗
kettle ['ket (ə) l]	壶	brush [brʌʃ]	刷子
sieve [sɪv]	面筛		

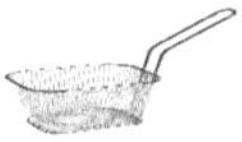

Expanded words & expressions（拓展单词和词组）

sift [sɪft]	筛	flatten ['flæt (ə) n]	弄平
simmer ['sɪmər]	焖	strain [streɪn]	过滤
stew [stjuː]	煨，慢炖	mix [mɪks]	混合，搅拌
fix.. to go with...	配上与……相配的……	fill it to top	装得很满
go ahead	干吧，做吧	first...and then	首先……然后

Lead in（导入）

Activity 1 Look and match.（看图并连线）

stockpot   frying pan   braising pan   sauce pan

Activity 2 Decide true or false.（判断正误）

(　　) 1. Frying pan is used to fry meat, fish, eggs, and etc.

(　　) 2. We use sauce pan to cook meat, fish, etc.

(　　) 3. The trainee can braise beef in a braising pan.

(　　) 4. We can't prepare the soup with the stockpot.

Activity 3 Oral practice.（口语练习）

A: Clean the ________, please. It is dirty.

B: No problem.

frying pan

sauce pan

braising pan

stockpot

Focus on learning（聚焦学习）

Part One

Activity 1 Read.（读词组）

sugar pan 糖锅　　grill pan 烤肉锅　　crepe pan 煎薄饼锅　　sauté pan 嫩炒锅

Activity 2 Look and write.（看图写英文）

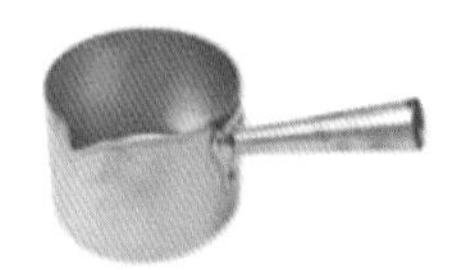

________　　________　　________　　________

Activity 3 Read and answer.（读对话，回答问题）

Questions: 1.What can you do with the frying pan?

2.How to cook the chicken in a sauce pan?

Mike : What's this?

Chef : It's a frying pan.

Mike : Oh, What can I do with it?

Chef : You can fry meat, fish, eggs, and etc.

Mike : What pot do I cook the chicken in?

Chef : Cook the chicken in a sauce pan.

Mike : What shall I do first?

Chef : Cook the chicken in butter. Then fix the sauce to go with the chicken.

Mike : In the same pan?

Chef : Right.

A: Mike, what can I do with the______ ?

B: You can ___________.

(frying pan/fry meat; sauce pan/make sauce; braising pan/braise beef; grill pan/grill meat)

Activity 4 Listen and complete.（听录音，完成对话）

A. Bring the water to boil	B. Put the hens in the stockpot
C. simmer them in the stockpot for one hour	D. Pour some water into the stockpot

Mike: Where shall I put the hens?

Chef:________________________.

Mike: What is the next?

Chef:______________________. Don’t fill it to the top.

Mike: And then?

Chef: ________________________. Then lower the fire.

Mike: To simmer?

Chef: Yes, ____________.

Activity 5 Pair work.（对话练习）

Mike is talking about the grill pan with Tom.

Mike

Tom

Part Two

Activity 1 Read.（读单词和词组）

soup ladle 汤匙	mixing bowl 拌菜碗
spider 漏勺	grater 擦菜器
whisk 打蛋器	fish poacher 蒸鱼盒
mallet 木槌	roasting tray 烤盘

Activity 2 Look, write and talk.（看图写英文，并进行对话练习）

A : Hand me the______, please.

B : Here you are.

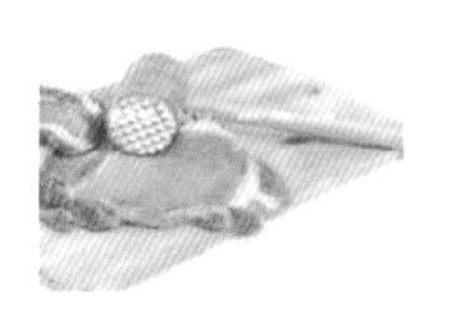

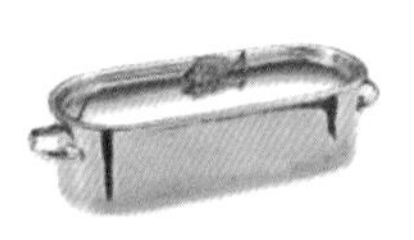

Activity 3 Complete and read.（完成并朗读对话）

A. go ahead	B. you will soon have mayonnaise
C. use a whisk	D. No, first add vinegar, salt and lemon

Mike : Shall I beat these eggs?

Chef : Yes, ____________________.

Mike : And then add oil?

Chef : _______________ , and then add oil.

Mike : And beat the ingredients with a whisk?

Chef : Right, ________________________.

Mike : Can I try it?

Chef : Of course, ____________________.

Part Three

Activity 1 Tick.（勾出你在厨房中经常使用的工具）

cutting board 切菜板　　cover 餐盘罩　　roasting fork 烤肉叉

kettle 壶　　pot rack 锅架　　frying basket 油炸篮

Activity 2 Look and write.（看图写英文）

 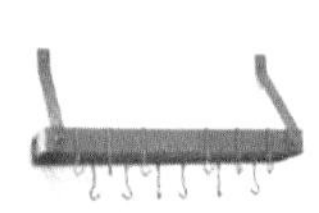    

______ ______ ______ ______ ______ ______

Activity 3 Decide true or false.（判断正误）

(　　) 1.Lift the meat with the roasting fork.

(　　) 2.We often cover the dish with a cover.

(　　) 3.Wooden cutting boards are cleaner than plastic.

(　　) 4.We can fry the potato chips with the frying basket.

Activity 4 Fill in and evaluate.（填空并自我评价）

1. st_ _k　　2. spid _ _　　3. gra_er　　4. si_ _er　　5. m_xing bowl

6. l_dle　　7. st_ck pot　　8. chi _ _en　　9. sau_e　　10. m_llet

Assessment : If you can write 8-10 words, you are perfect.

If you can write 4-7 words, you are good.

If you can write 1-3 words, you will try again.

Give it off（知识拓展）

Practice 1 Match.（连线）

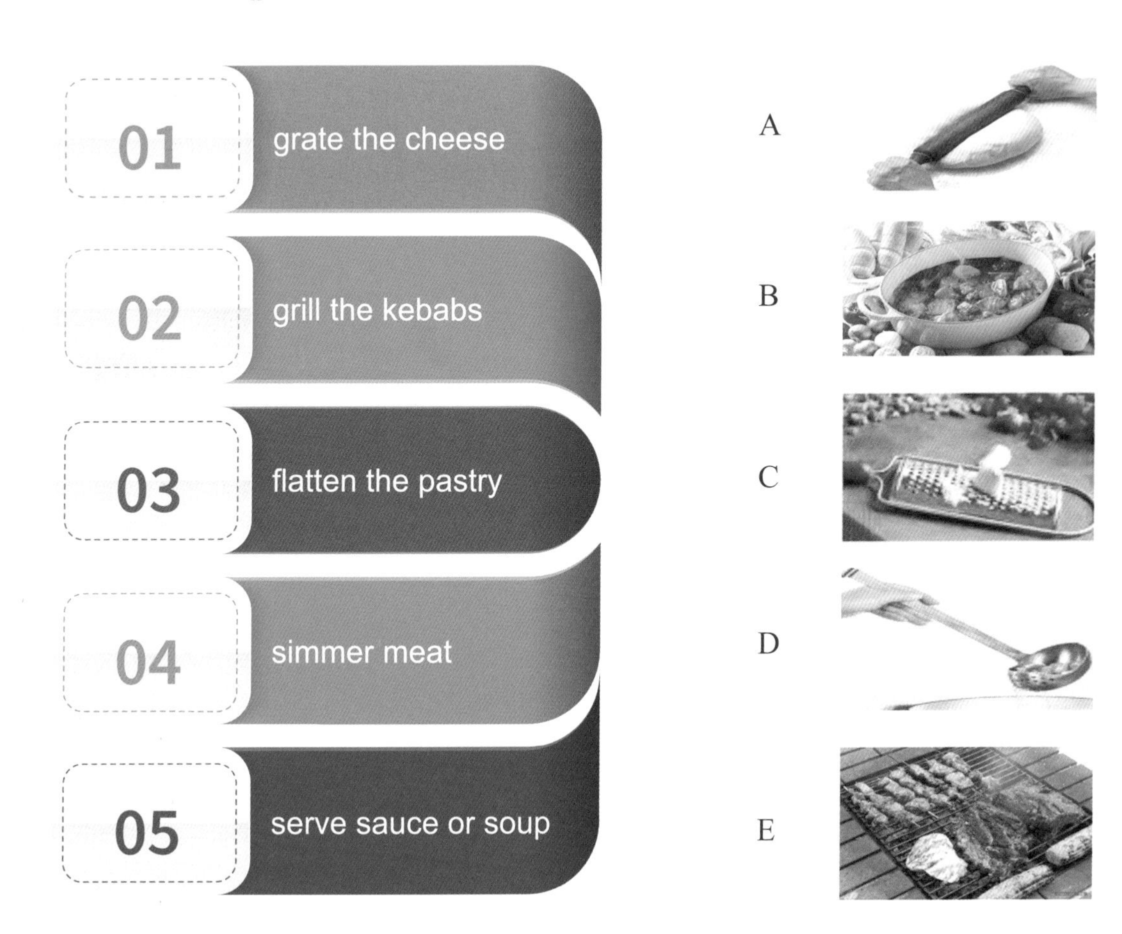

Practice 2 Summarize.（总结本单元的重点句型）

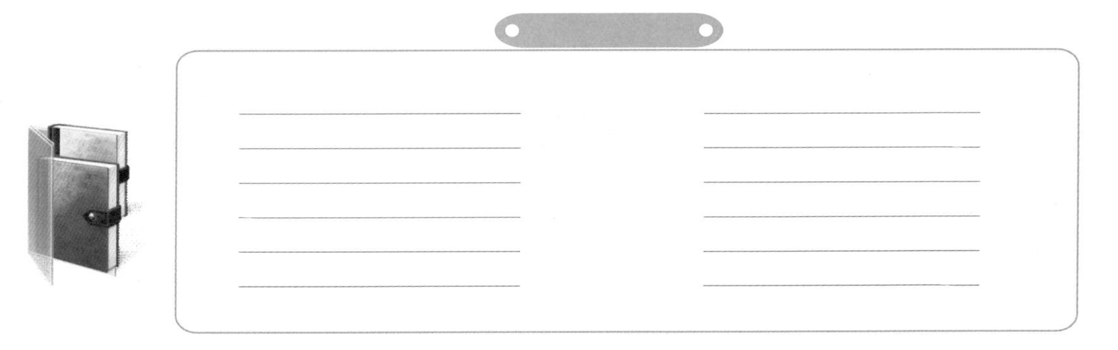

Practice 3 Make dialogues.（对话练习）

A: How do you say this in English?

B: We call it *a serving spoon*.

A: What are we going to do?

B: We're going to *serve potatoes with it*.

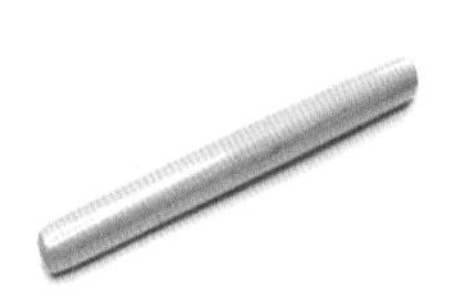

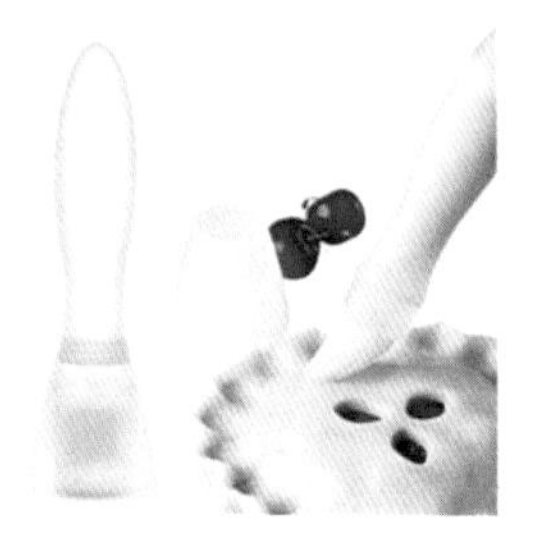

This is called *a whisk*.

It is used for *beating eggs*.

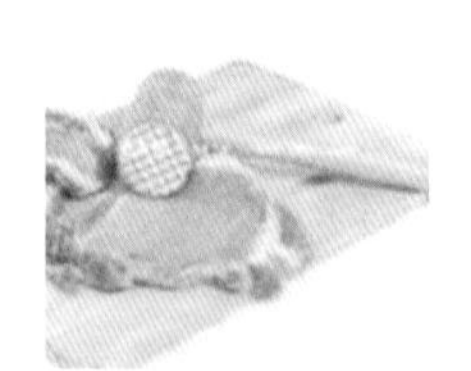

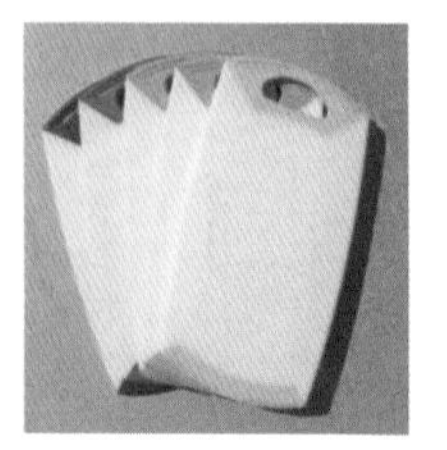

Unit exercises（单元练习）

Exercise 1 Choose the best answers.（选择正确答案）

1. The vegetable chef needs a _____. He wants to chop up the onions.

   A. whisk　B. sieve　C. spider　D. cutting board

2. We usually use the ______ to sift flour.

   A. mixing bowl　B. sieve　C. colander　D. frying basket

3. The chef will put the chicken on the ________and then put the tray into the oven.

   A. spider　B. spoon　C. ladle　D. roasting tray

4. I want to make an apple pie. Let's flatten it with a ______.

   A. mixing bowl　B. spoon　C. whisk　D. rolling pin

5. —How shall I serve the carrots?

   —You can use a ______.

   A. ladle　B. sponge　C. serving spoon　D. grater

6. The chef is mixing the salad in this______.

   A. mixing bowl　B. spider　C. spoon　D. ladle

7. The chicken soup is ready. Please serve them with the_____.

   A. sponge　B. serving spoon　C. ladle　D. skimmer

8. —What shall I do with it?

   —Use a _______to flatten meat.

   A. mallet　B. spoon　C. whisk　D. rolling pin

9. The pastry chef wanted to make a pizza. He asked the trainee to grate the cheese with the______.

   A. whisk　B. sieve　C. spider　D. grater

10. If you want to make potato chips, you should put them into the _____.

    A. mallet　B. spoon　C. frying basket　D. whisk

Exercise 2 Make sentences in right order.（连词成句）

1. is used to, the, frying pan, meat, fish, fry, etc

____.

2. need, I, cutting board, a, meat, to, cut

____.

3. flatten, going, am, to, I, meat, a, mallet, with, the

____.

4. make, omelette, I, to, frying pan, want, in, the

____.

5. butter, he, is, the, melt, going, sauce pan, to, in

____.

Exercise 3 Fill in the blanks.（填空）

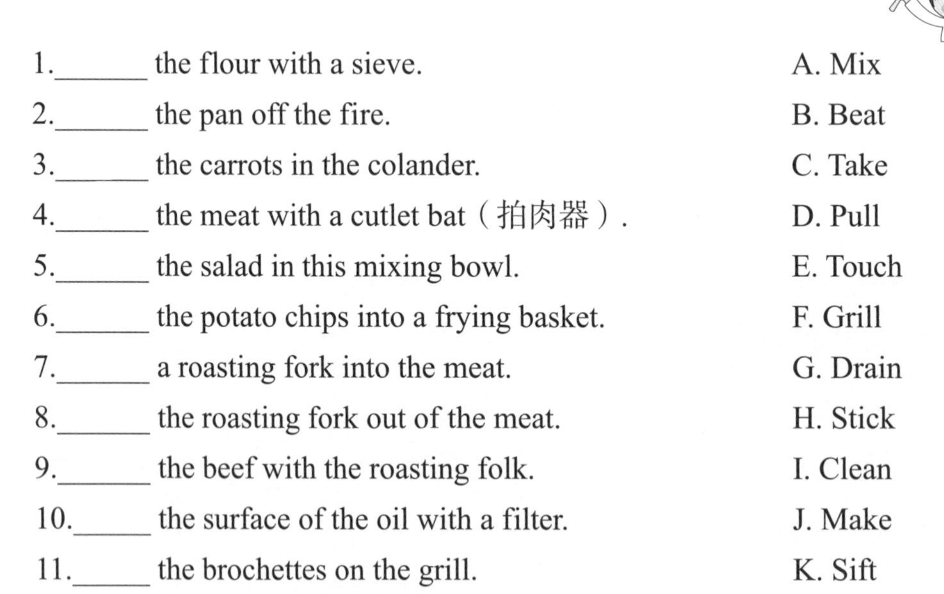

1.______ the flour with a sieve.	A. Mix
2.______ the pan off the fire.	B. Beat
3.______ the carrots in the colander.	C. Take
4.______ the meat with a cutlet bat（拍肉器）.	D. Pull
5.______ the salad in this mixing bowl.	E. Touch
6.______ the potato chips into a frying basket.	F. Grill
7.______ a roasting fork into the meat.	G. Drain
8.______ the roasting fork out of the meat.	H. Stick
9.______ the beef with the roasting folk.	I. Clean
10.______ the surface of the oil with a filter.	J. Make
11.______ the brochettes on the grill.	K. Sift
12.______ shish kebabs with skewers（烤肉叉子）.	L. Put

Exercise 4 Translate into Chinese or English.（中英文互译）

1. 用滤网筛面粉。

____

2. 用刷子在面饼上刷蛋黄液。

____

3. Clean the surface of the oil with a skimmer.

____

4.—How shall I cook these potato chips?

—Put the potato chips in a frying basket.

____

5. —What shall I do with the carrots?

—To drain them.

____

Learning tips（学习提示）

Cutting boards have many colors, and different colors are for different usages, so match the right pictures with their usages.

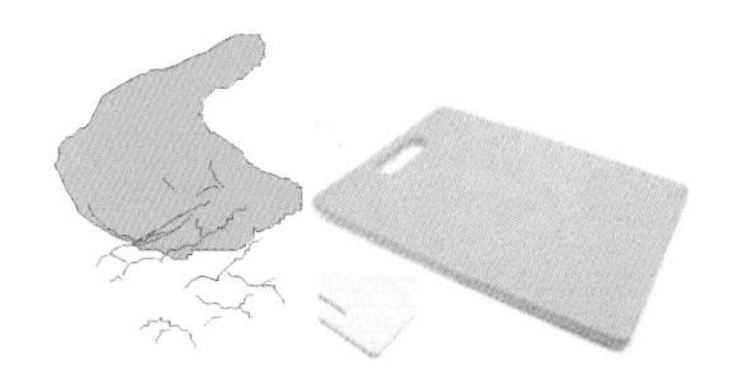

A.Blue for fish.

B. Red for meat.

C.Yellow for poultry.

D.Green for vegetables.

Culture knowledge（文化知识）

超实用的厨房用具，分分钟震撼你！

做美食必备的复古计时器

厨房中有一个计时器，不仅能够克服忘性，而是可以保证美食烹饪过程的时间精准度。计时器使用机械发条，搭配上精细的刻度，计时更加精准。

一招搞定切鸡蛋的鸡蛋分割器

用刀切鸡蛋？当然可以，但是一刀一刀切多累，还怕切得不均匀。一把锌合金的鸡蛋分割器，一秒搞定你的鸡蛋切割哦！这不仅省事，而且切刀间隔排布均匀，分切整齐，每一片大小一致，外形美观，适宜摆盘，让你有大厨的即视感！

三秒完成苹果去核切瓣的切割器

水果切分器可以切苹果、雪梨、火龙果，有效去核，平均等分，操作很简单，中间的孔对准水果的中心，两手握着手柄向下压，分成等份，就成功了！采用加厚的锌合金打造，304不锈钢刀片，也很容易清洗，一冲即可，简单不费事。

Unit 4

Kitchen Knives

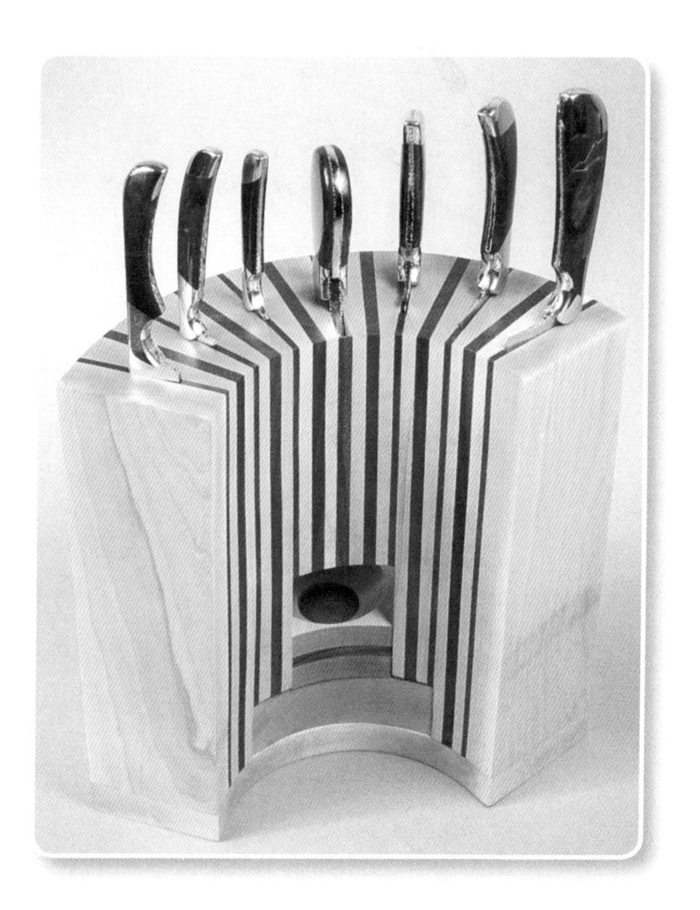

Learning goals（学习目标）

You will be able to

★ tell the names of different kinds of knives.

★ get familiar with the usages of different kinds of knives.

★ make short conversations on talking about knives.

Word list（单词表）

Key words & expressions（重点单词和词组）

paring ['peərɪŋ] knife	水果刀	spatula ['spætʃələ]	刮铲
oyster ['ɔɪstə] knife	牡蛎刀	chopping ['tʃɒpɪŋ] knife	砍刀
chef's [ʃefs] knife	厨刀	cheese knife	奶酪刀
salmon ['sæmən] knife	三文鱼刀	boning ['bəʊnɪŋ] knife	剔骨刀
whetstone ['wetstəʊn]	磨刀石	sharpening ['ʃɑːpənɪŋ] steel	磨刀棒
knife sharpener ['ʃɑːpnə (r)]	磨刀器	electric [ɪ'lektrɪk] knife sharpener	电动磨刀器
carving ['kɑːvɪŋ] knife	雕刻刀	fish scissors ['sɪzəz]	鱼剪
pizza cutter ['kʌtə]	比萨切刀	bread [bred] knife	面包刀
roasting ['rəʊstɪŋ] fork	烤叉	peeler ['piːlə]	削皮器

Expanded words & expressions（拓展单词和词组）

chop [tʃɒp]	剁，砍，劈	spread [spred]	摊，抹
mince [mɪns]	切碎	dull [dʌl]	钝的，不锋利的
dice [daɪs]	切丁	crush [krʌʃ]	压碎，碾压
slice [slaɪs]	将……切成薄片	cut up	切，割
electric bone saw	电动骨锯	sharpen the knife	磨刀

Lead in（导入）

Activity 1 Look and match.（看图并连线）

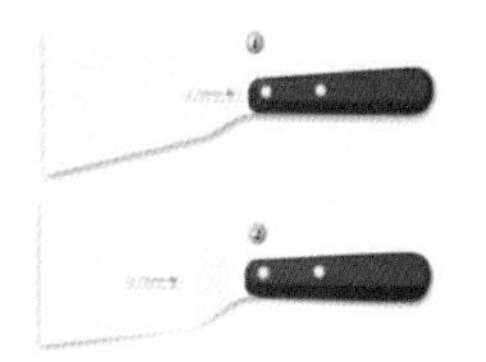  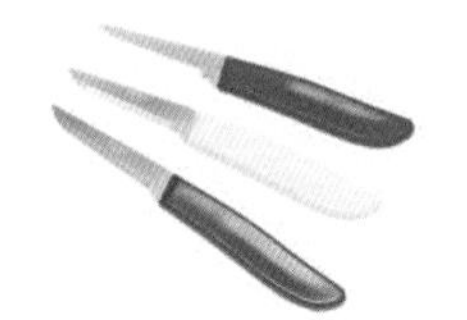 

paring knife	spatula	oyster knife	chopping knife

Activity 2 Write.（根据描述写出刀具的英文表达）

1. ____________ is used to mix, lift and spread materials.
2. ____________ is a knife for chopping or mincing meat, vegetables, etc.
3. We use an ___________ to open oysters.
4. We cut and peel fruits with a ___________.

Activity 3 Oral practice.（口语练习）

A: Can you hand me a________?
B:Sure. Here you are.
Sorry, I don't have one.

paring knife
chopping knife
oyster knife
spatula

Focus on learning（聚焦学习）

Part One

Activity 1 Read.（读词组）

chef's knife 厨刀	cheese knife 奶酪刀
salmon knife 三文鱼刀	boning knife 剔骨刀

Activity 2 Look and write.（看图写英文）

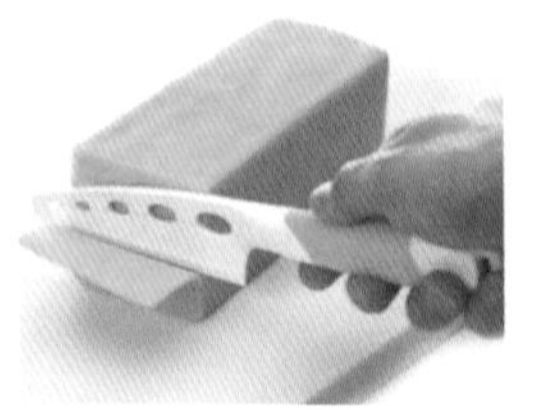

__________ __________ __________ __________

Activity 3 Read and answer.（读对话，回答问题）

Questions: 1. Is the chef's knife for special jobs?
2. What is Mike going to cut?

Mike : Is my chef's knife for special jobs?
Chef: No. Your chef's knife is for many different things.
Mike: Can I cut up the meat with my chef's knife?
Chef: Are there any bones in the meat?
Mike: No.
Chef: Then go ahead.
Mike: Now I want to cut up some cheese.
Chef: You may use your cheese knife.

A: I need to/want to cut up (slice, open) some __________.
B: Use your__________.
(cheese/cheese knife; salmon/salmon knife; onion/chef's knife; oysters/oyster knife)

Activity 4 Listen and complete.（听录音，完成对话）

A. how to use it	B. use the electric bone saw
C. to cut it into two	D. to cut big bones

Mike: Oh, this is a big bone. I want ______________.

Chef: Don't use your chef's knife.

Mike: What shall I use then?

Chef: You can _______. We never use chef's knives ______________.

Mike: Why?

Chef: It's dangerous. We always use an electric bone saw.

Mike: Oh, thank you. Can you show me _______?

Chef: Sure.

Activity 5 Pair work.（对话练习）

eg Mike is talking about his chef's knife with Tom.

Mike

Tom

Part Two

Activity 1 Read.（读单词和词组）

whetstone 磨刀石	sharpening steel 磨刀棒
electric knife sharpener 电动磨刀器	knife sharpener 磨刀器

Activity 2 Look, write and talk.（看图片，写出加工工具的英文表达并进行句型练习）

A : What shall I use to sharpen the knife?

B : You may use ____________.

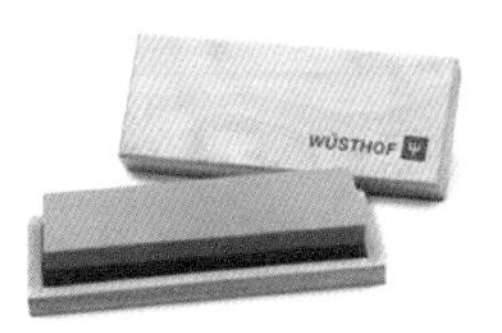

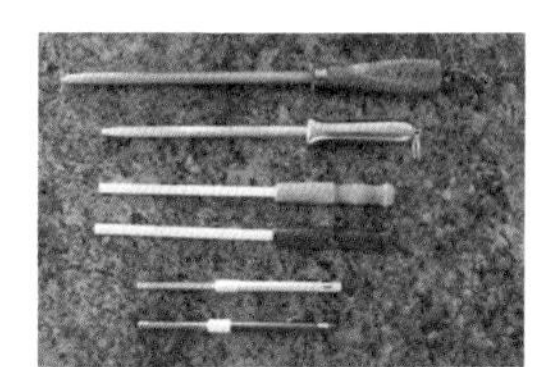

____________　____________　____________　____________

Activity 3 Complete and read.（完成并朗读对话）

A. use a whetstone	B. My knife is dull
C. You'd better sharpen it	D. you can use my steel

Mike : __________. I can't cut the beef.

Chef : ________________________.

Mike : Where can I find a sharpening steel?

Chef : Here, ____________________________.

Mike : Thank you. What else can I use to sharpen my knife?

Chef : You may ___________ , or you can also use an electric knife sharpener, which is safer.

Mike : Oh, I see. Thank you.

Part Three

Activity 1 Tick.（勾出你在厨房中从未用过的工具）

oyster knife	spatula	paring knife
chopping knife	cheese knife	salmon knife

Activity 2 Look and write.（看图写英文）

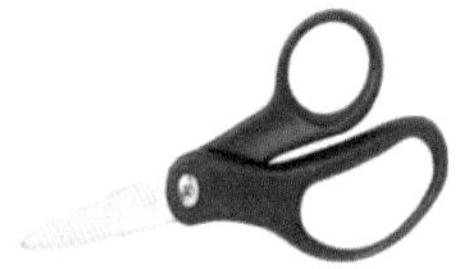

______ ______ ______

______ ______ ______

Activity 3 Fill in and evaluate.（填空并自我评价）

1. d_ _l 2. ch_ _se 3. b_ne 4. sp_c_a_ 5. s_ _m_n
6. s_ _t_l_ 7. el_ _t_ _c 8. s_fe 9. p_ _ing knife 10. o_st_ _

Assessment: If you can write 8-10 words, you are perfect.
If you can write 4-7 words, you are good.
If you can write 1-3 words, you will try again.

Give it off（知识拓展）

Practice 1 Match.（连线）

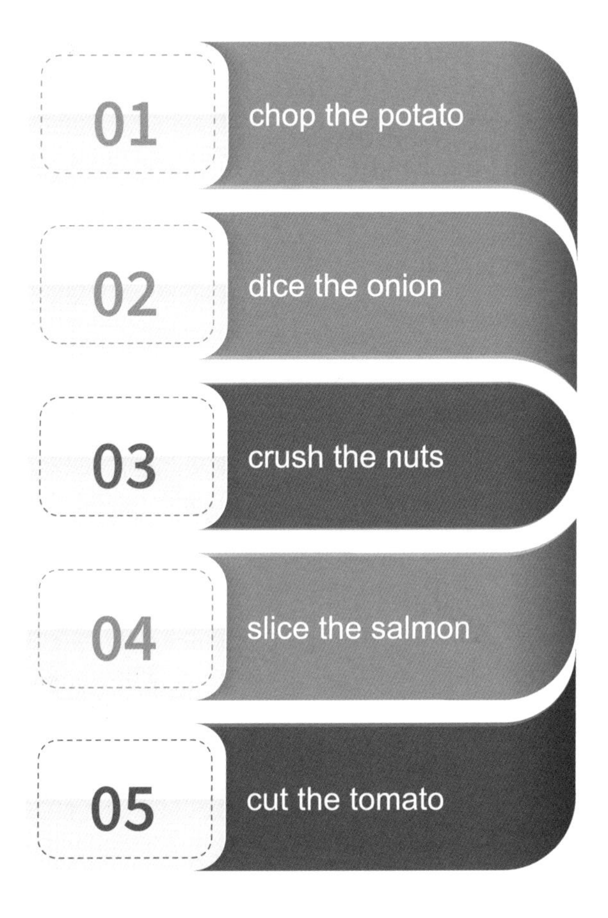

A

B

C

D

E

Practice 2 Summarize.（总结本单元的重点句型）

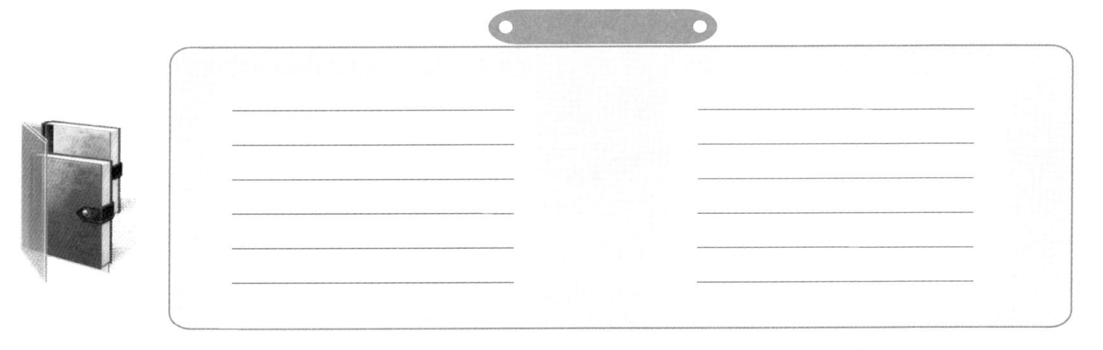

Practice 3 Make dialogue.（对话练习）

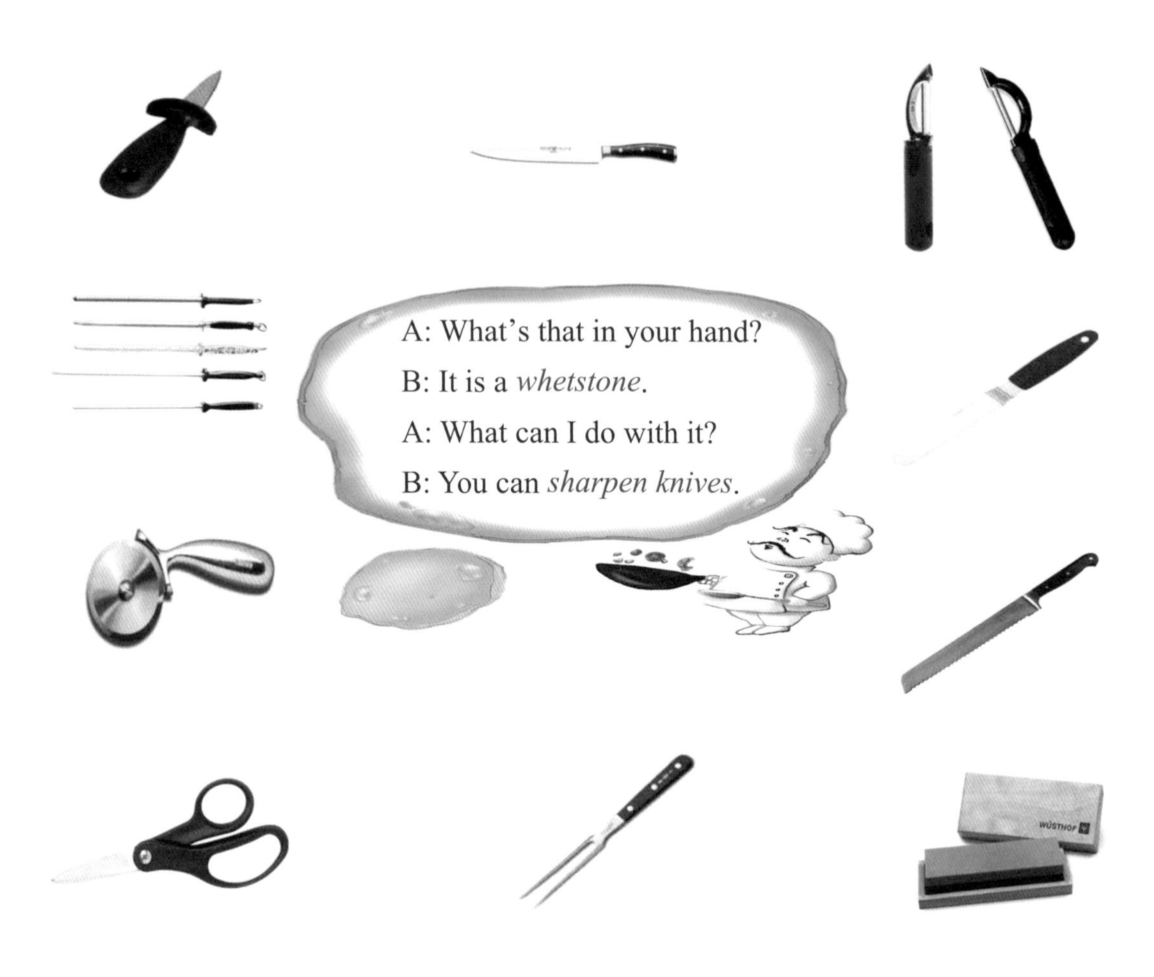

eg

This is a *chopping knife*.
I am going to *chop up the beef with it*.

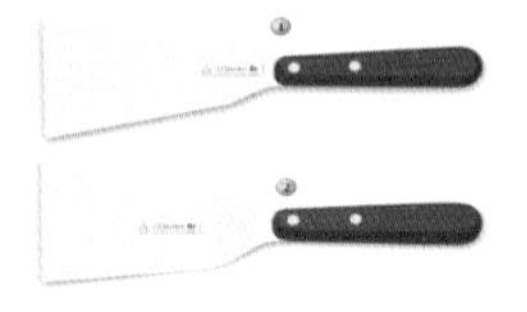

Unit exercises（单元练习）

Exercise 1 Decide true or false.（判断正误）

(　　) 1. Pizza cutter is used to cut pizza.

(　　) 2. We use bread knife to cut cheese.

(　　) 3. We can lift meat with a roasting fork.

(　　) 4. Carving knife is used to peel fruits.

(　　) 5. We can use oyster knife to cut different kinds of food.

Exercise 2 Make sentences in right order.（连词成句）

1. is, to, used, a, paring knife, peel, fruits

______________________________________.

2. cut up, things, chef's knife, can, many

______________________________________.

3.safer, the electric bone saw, much, is

______________________________________.

4. funny, the oyster knife, looks, very

______________________________________.

5. vegetables, is, the chef, cutting, some

______________________________________.

Exercise 3 Write and make sentences.（根据图片写出英文，并进行句子练习）

We _*peel potatoes*_ with _*a peeler*_.

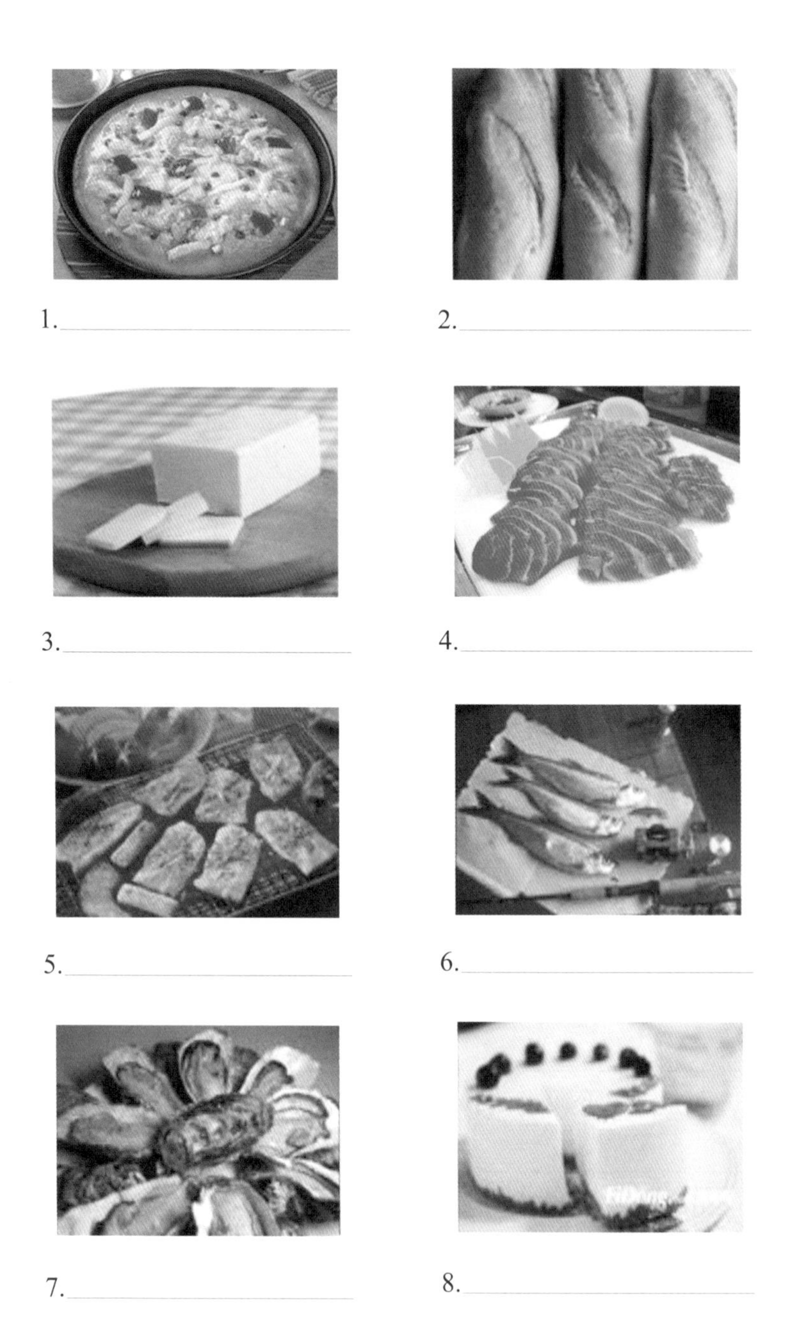

1. ______________ 2. ______________

3. ______________ 4. ______________

5. ______________ 6. ______________

7. ______________ 8. ______________

Exercise 4 Translate into Chinese or English.（中英文互译）

1. 请用你的厨刀把胡萝卜切成小块。

____

2. —我用什么削这些土豆？
   —你可以用削皮器，给你用这个吧。

   ____

3. A boning knife is used to remove bones from meat.

   ____

4. You can cut many different things with kitchen knives, so they are very useful.

   ____

5. Use the fish scissors to cut the fish, please.

   ____

Exercise 5 Make a summary.（用中文或英文总结）

课本上出现的刀具

____
____
____
____
____
____

我还知道的刀具

____
____
____
____
____
____

Learning tips（学习提示）

Do you know what you should pay attention to when using knives？

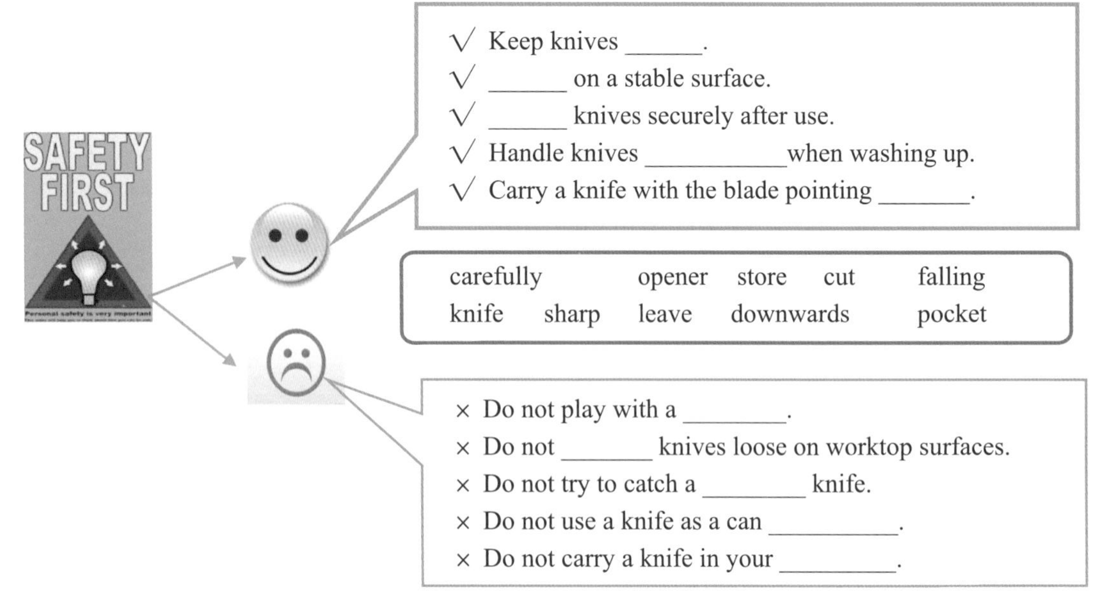

Culture knowledge（文化知识）

厨房刀具使用注意事项

1. 刀具摆放在正确安全的位置，必须平放，不宜放在操作台边沿及过高处。
2. 使用前注意检查卫生、破损、变形等情况。
3. 根据刀具种类进行正常加工，避免一刀多用的情况。
4. 临时放置刀具时，将把手朝向自己，禁止用抹布等遮挡住刀具。
5. 操作（用刀）时，其他人不能影响（如碰撞、打闹、聊天等）。
6. 厨房刀具严禁作为非工作用途工具使用。
7. 所有厨房刀具任何人未经许可不能私自带出店外。
8. 厨房刀具根据使用情况由具体使用人维护保养。

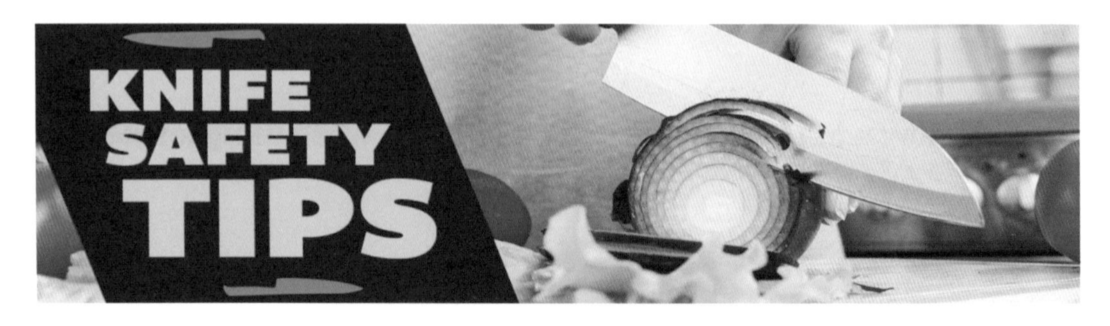

Unit 5

Condiments & Spices

Learning goals（学习目标）

You will be able to

★ know the English names of condiments and spices.

★ know how to classify condiments and spices.

★ talk about the taste of condiments.

Word list（单词表）

Key words & expressions（重点单词和词组）

condiment ['kɒndɪmənt]	调味品
vinegar ['vɪnɪgə(r)]	醋
ketchup ['ketʃəp]	番茄酱
sugar ['ʃʊgə (r)]	糖
chicken powder ['paʊdə (r)]	鸡精
mustard ['mʌstəd]	芥末
pepper powder	胡椒粉
salad oil ['sæləd ɔɪl]	植物油
cinnamon ['sɪnəmən]	桂皮
rosemary ['rəʊzməri]	迷迭香
thyme [taɪm]	百里香
bay leaf ['beɪ liːf]	香叶
nutmeg ['nʌtmeg]	肉豆蔻
marjoram ['mɑːdʒərəm]	牛膝草
mint [mɪnt]	薄荷
spice [spaɪs]	（调味）香料
honey ['hʌni]	蜂蜜
chili oil ['tʃɪli ɔɪl]	辣椒油
salt [sɔːlt]	食盐
soy sauce [ˌsɔɪ 'sɔːs]	酱油
jam [dʒæm]	果酱
olive oil [ˌɒlɪv 'ɔɪl]	橄榄油
butter ['bʌtə (r)]	黄油
clove [kləʊv]	丁香
fennel ['fenl]	茴香，小茴香
dill [dɪl]	莳萝
saffron ['sæfrən]	番红花
sage [seɪdʒ]	鼠尾草
tarragon ['tærəgən]	他力根香草
basil ['bæzl]	罗勒

Expanded words & expressions（拓展单词和词组）

sour ['saʊə (r)]	酸的	sweet [swiːt]	甜的
bitter ['bɪtə (r)]	苦的	salty ['sɔːlti]	咸的
spicy ['spaɪsi]	辣的	flavor ['fleɪvə]	风味，滋味
beverage ['bevərɪdʒ]	饮料	for exampl [ɪg'zɑːmpl]	例如
cereal ['sɪəriəl]	麦片粥	substance ['sʌbstəns]	物质
pickled ['pɪkld]	腌渍的	compound ['kɒmpaʊnd]	混合的
seasoning ['siːznɪŋ]	调味品，加调料	black pepper	黑胡椒
pungent ['pʌndʒənt]	刺鼻的	culinary ['kʌlɪnəri]	烹饪的
nutty ['nʌti]	坚果味的	fragrant ['freɪgrənt]	芳香的
aromatic [ˌærə'mætɪk]	芳香的，有香味的		

Lead in（导入）

Activity 1 Look and match.（看图并连线）

vinegar　　honey　　ketchup　　chili oil　　sugar　　salt

Activity 2 Match.（中英文配对）

A	B
1. sour	a. 咸的
2. sweet	b. 苦的
3. bitter	c. 辣的
4. salty	d. 酸的
5. spicy	e. 甜的

Activity 3 Oral practice.（口语练习）

A : What is the flavor of ______?

B : It is ____________.

salt/salty

honey/sweet

vinegar/sour

sugar/sweet

......

Focus on learning（聚焦学习）

Part One

Activity 1 Read.（读单词和词组）

chicken powder 鸡精	soy sauce 酱油	mustard 芥末	jam 果酱
pepper powder 胡椒粉	olive oil 橄榄油	salad oil 植物油	butter 黄油

Activity 2 Look and write.（看图写英文）

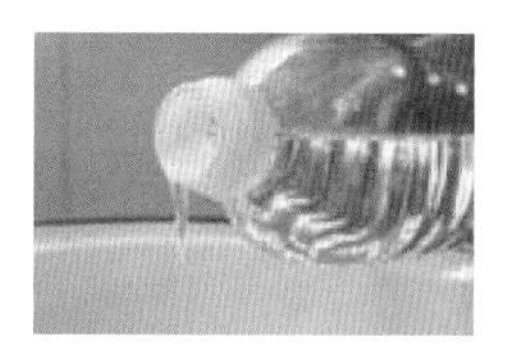   

__________ __________ __________ __________

__________ __________ __________ __________

Activity 3 Read and answer.（读对话，回答问题）

Questions: 1. Is sugar very sweet?
2. What is it used for?

Mike : What shall I call it in English? Is it sour?
Chef : It is sugar. It is very sweet.
Mike : What is it used for?
Chef：It can be put in some beverage.
Mike : For example?
Chef : Such as coffee, milk, tea or lemon juice.
Mike : Oh, I see. Thank you.

 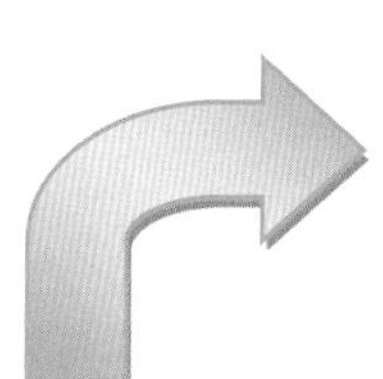

A: Mike, what is the flavor of_____ ? (salt/sugar...)
B: It is _________. (salty/sweet...)
A: Is it sour?
B: No, it is_________. (salty/sweet/hot...)

Activity 4 Listen and complete.（听录音，完成对话）

A. gives food a strong flavor

B. compound sauce and so on

C. How many kinds of condiments are there

D. five types of condiments

Mike : What is condiment?

Chef : It is a substance that ________________.

Mike : ____________________________?

Chef : There are ________________.

Mike : What are they?

Chef : They are salty, sweet, spicy, pickled condiments, ________________.

Activity 5 Pair work.（对话练习）

eg Mike is talking about olive oil with Tom.

Mike

Tom

Part Two

Activity 1 Read.（读单词和词组）

cinnamon 桂皮	clove 丁香	rosemary 迷迭香	fennel 茴香，小茴香
thyme 百里香	dill 莳萝	bay leaf 香叶	saffron 番红花

Activity 2 Look, write and talk.（看图写英文，并进行对话练习）

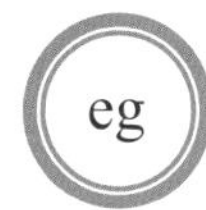

A : What do you want me to do?

B : Hand me the__________.

__________ __________ __________ __________

   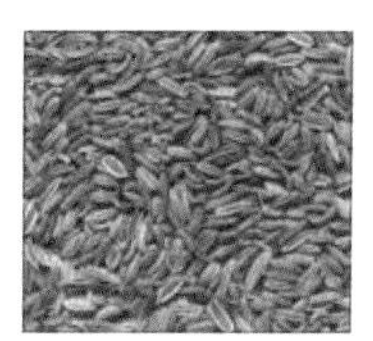

__________ __________ __________ __________

Activity 3 Complete and read.（完成并朗读对话）

A. It is slightly bitter and strongly spicy	B. Yes, you're right
C. It's used for seasoning	D. It is called bay leaf

Mike : What shall I call it in English?

Chef : ________________________.

Mike : What's the flavor of it?

Chef : ____________________.

Mike : What is it used for?

Chef : __________________.

Mike : Is this common in the kitchen?

Chef : ___________________.

Part Three

Activity 1 Tick.（勾出你熟悉的香料）

nutmeg 肉豆蔻	sage 鼠尾草	marjoram 牛膝草
tarragon 他力根香草	mint 薄荷	basil 罗勒

Activity 2 Look and write.（看图写英文）

______ ______ ______ ______ ______ ______

Activity 3 Decide true or false.（判断正误）

(　　) 1. Nutmeg is brown and round, it has a strong flavor.
(　　) 2. Cloves are condiment, not spice.
(　　) 3. Black pepper is less pungent.
(　　) 4. Sage is often used in cooking meat and vegetable soup.

Activity 4 Fill in and evaluate.（填空并自我评价）

1. s_lt	2. sug_r	3. s_lty	4. must_rd	5. bay le_f
6. sw_ _t	7. bi_ _er	8. pe_per	9. di_ _	10. thym_

Assessment : If you can write 8-10 words, you are perfect.
If you can write 4-7 words, you are good.
If you can write 1-3 words, you will try again.

Give it off（知识拓展）

Practice 1 Match.（连线）

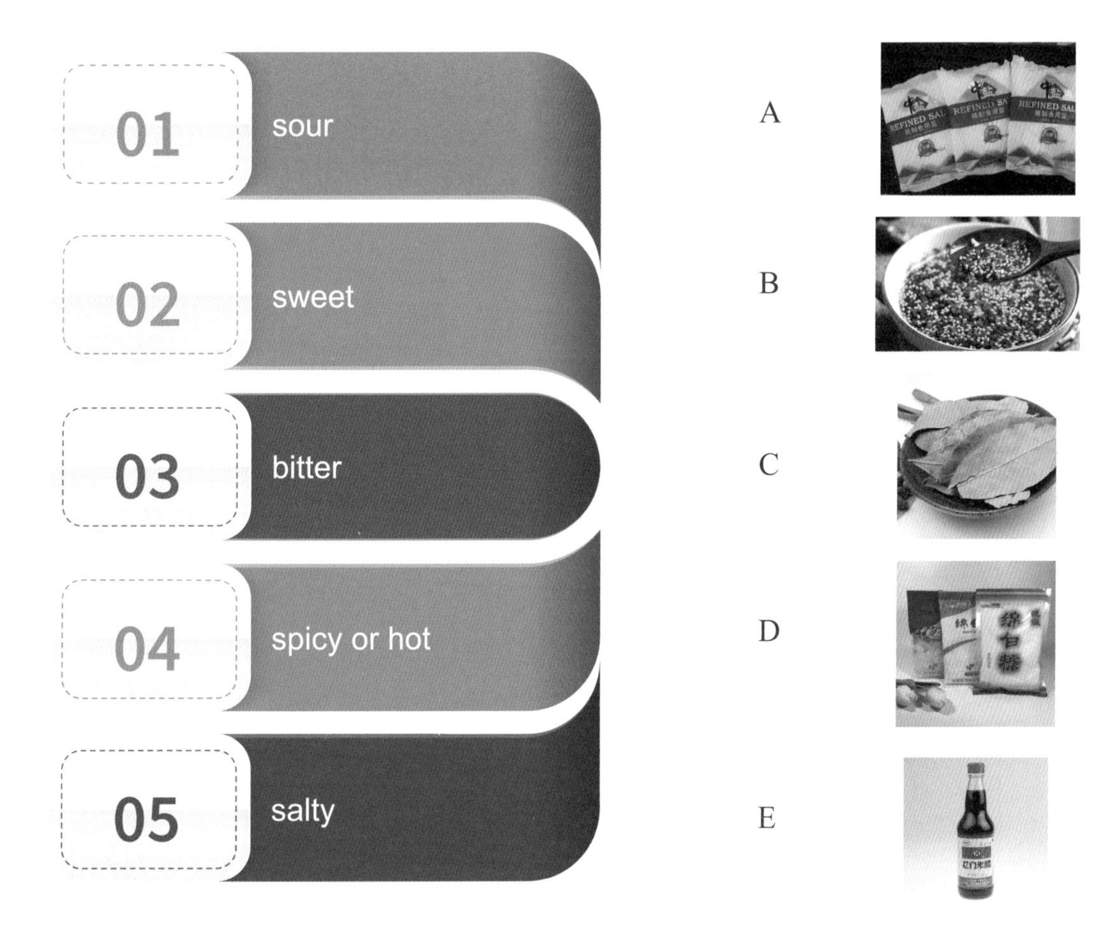

Practice 2 Summarize.（总结本单元的重点句型）

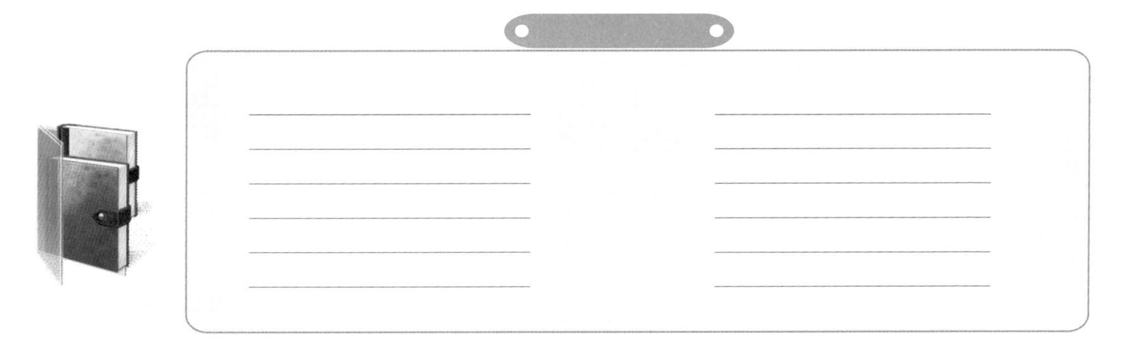

Practice 3 Make dialogues.（对话练习）

A: Can you tell me how do you say it in English?

B: It is *sugar*.

A: What is the culinary use of it?

B: It is used *as seasoning*.

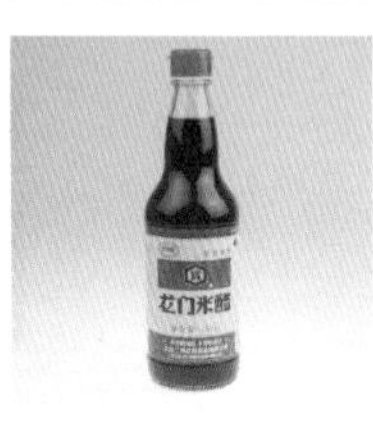

This is *sugar*.
It is *sweet*.

This is ________.
It is ________.

That is ________.
It is ________.

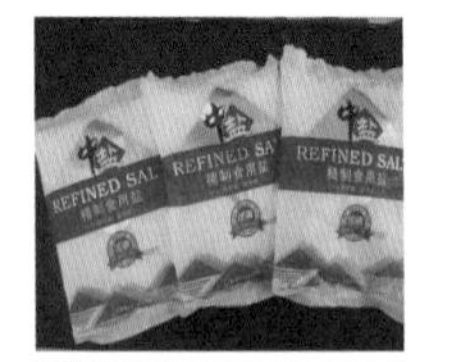

They are ________.
They are ________.

We call it ________.
It is ________.

Unit exercises（单元练习）

Exercise 1 Choose the best answers.（选择最佳答案）

1. Nutmeg is nutty and slightly ________.

   A. spicy B. sour C. bitter D. sweet

2. Clove is sweetly pungent and _______ fragrant.

   A. strongly B. little C. few D. much

3. Cinnamon is _______ and aromatic.

   A. cold B. warm C. cool D. hot

4. There are_______types of condiments.

   A. 3 B. 4 C. 5 D. 6

5. Bay leaf is slightly ______ and strongly spicy.

   A. bitter B. sweet C. hot D. salty

Exercise 2 Match.（单词与词义配对）

A	B
nutmeg	contributes a yellow-orange coloring to food
fennel	is often used in cooking meat and vegetable soup
sage	is not only nutty, but also warm and slightly sweet
saffron	is often used in egg, fish and other dishes

Exercise 3 Fill in and learn.（填空并学习醋的分类）

balsamic _______ 意大利香脂醋　　champagne_____ 香槟醋

________vinegar 他力根香醋　　__________ vinegar 葡萄酒醋

_______ vinegar 苹果醋　　___________vinegar 白醋

Exercise 4 Make sentences in right order.（连词成句）

1. five, are, there, condiments, of, types

______________________________________.

2. of, what, the, flavor, is, it

______________________________________?

3. sugar, sweet, is,very

______________________________________.

4. is, cooking, often, in, used, soup, meat and vegetable, sage

______________________________________.

5. in, shall, I, call, it, English, what

______________________________________?

Exercise 5 Translate into Chinese or English.（中英文互译）

1. What is the flavor of sugar?

____

2. Bay leaf is slightly bitter and strongly spicy.

____

3. Condiment gives food a strong flavor.

____

4. 调味品共有五类。

____

5. 鼠尾草常用于烹调肉类和蔬菜汤。

____

Learning tips（学习提示）

Brandy, whisky, gin, red wine,white wine, champagne, which could help to get rid of the unpleasant smell are often used as condiments when cooking, especially when meat or fish are cooked.

Culture knowledge（文化知识）

香料在西餐中的作用非常重要，用途非常广，腌肉、调沙拉汁、做汤、做甜点等都会用到，下面就给大家介绍一下西餐中比较常见的香料。

胡椒（Pepper），西餐中常用它来去腥和增味提香。整粒胡椒用在肉类、汤类、鱼类及腌渍类食品的调味和防腐中，白胡椒是成熟的胡椒种子脱皮加工而成，黑胡椒是未成熟的种子晒干后加工而成的。

丁香（Clove），很好认的一个香料，长得就像一枚小钉子。西餐中烹调肉类、腌泡菜、烘焙糕点以及调制甜酒时经常加入丁香，美国人喜欢用来洒在烧烤类食物上。中国菜里也会用到丁香，常用的十三香里面就有它。

肉桂（Cinnamon），长得像中餐常用的桂皮，但西方常用的品种是锡兰肉桂，气味香甜而清淡，非常适用于甜味浓郁的菜肴，甜品或鸡尾酒里会用到，加到肉桂苹果派、肉桂卷面包里都挺美味。

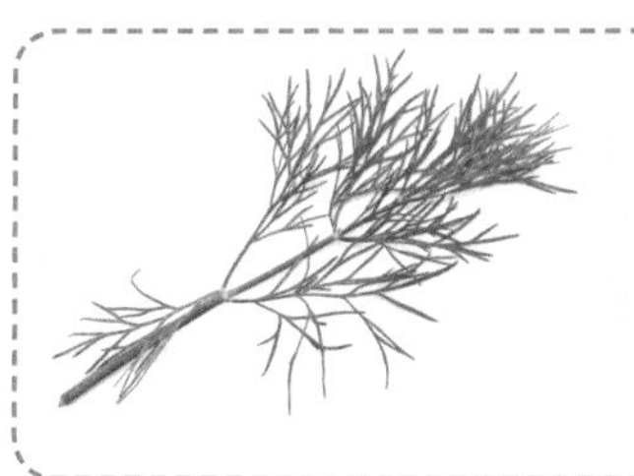

莳萝（Dill），长得像茴香但味道不同，略带辛辣味，可以鲜食，也可以干制成香料，打碎可用于制作酱料，也可以撒在鱼类冷盘上（烟熏三文鱼上常见）去腥添香做装饰，也可加入泡菜和汤品中。

Unit 6

Fruits & Nuts

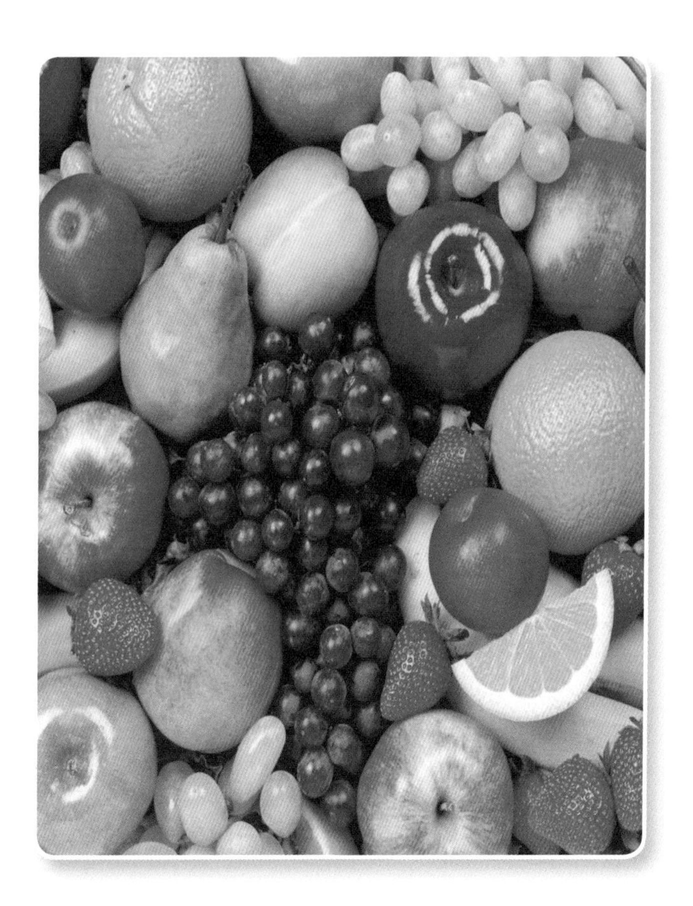

Learning goals（学习目标）

You will be able to

★ get familiar with the fruits and nuts.

★ describe the fruits with different color.

★ be familiar with the ways of making fruit dishes.

Word list（单词表）

Key words & expressions（重点单词和词组）

pear [peə (r)]	梨	
cherry ['tʃeri]	樱桃	
strawberry ['strɔːbəri]	草莓	
grape [greɪp]	葡萄	
grapefruit ['greɪpfruːt]	西柚	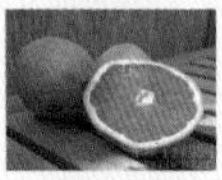
apricot ['eɪprɪkɒt]	杏	
pomegranate ['pɒmɪgrænɪt]	石榴	
kiwi ['kiːwi] fruit	猕猴桃	
lychee ['laɪtʃi]	荔枝	
star [stɑː(r)] fruit	杨桃	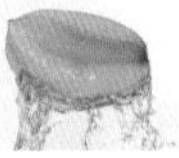
mango ['mæŋgəʊ]	芒果	
walnut ['wɔːlnʌt]	核桃	
chestnut ['tʃesnʌt]	栗子	
almond ['ɑːmənd]	杏仁	
honey melon ['melən]	哈密瓜	
watermelon ['wɔːtəmelən]	西瓜	
peach [piːtʃ]	桃子	
blueberry ['bluːbəri]	蓝莓	
plum [plʌm]	李子	
Chinese date [deɪt]	大枣	
lemon ['lemən]	柠檬	
pineapple ['paɪnæpl]	菠萝	
papaya [pə'paɪə]	木瓜	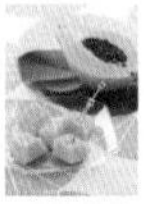
dragon ['drægən] fruit	火龙果	
coconut ['kəʊkənʌt]	椰子	
peanut ['piːnʌt]	花生	
hazelnut ['heɪzlnʌt]	榛子	
cashew ['kæʃuː]	腰果	

Expanded words & expressions（拓展单词和词组）

soak [səʊk]	浸泡	peel [pi:l]	剥皮
remove seeds [rɪ'mu:v] [si:dz]	去籽儿	ingredient [ɪn'gri:diənt]	原料
make salad ['sæləd]	做沙拉	teaspoon ['ti:spu:n]	茶勺
yogurt ['jɒgət]	酸奶	mix [mɪks] up	搅拌

Lead in（导入）

Activity 1 Look and match.（看图连线）

pear　honey melon　cherry　watermelon　strawberry　peach

Activity 2 Write.（写出下列水果颜色的英文表达）

cherry___________　peach___________　watermelon___________

strawberry___________　pear___________　honey melon___________

Activity 3 Oral practice.（口语练习）

A: What's your favorite fruit?

B: My favorite fruit is ___________.

cherry

strawberry

peach

honey melon

watermelon

Focus on learning（聚焦学习）

Part One

Activity 1 Read.（读单词和词组）

grape 葡萄	blueberry 蓝莓	grapefruit 西柚	plum 李子
apricot 杏	Chinese date 大枣	pomegranate 石榴	lemon 柠檬

Activity 2 Look and write.（看图写英文）

 ___________

 ___________

 ___________

 ___________

 ___________

 ___________

 ___________

 ___________

Activity 3 Read and answer.（读对话，回答问题）

Questions: 1. What shall we do today?

2. Where is he going to put the grapes?

Mike: What shall we do with the grapes?

Chef: Wash and soak them in the salt water.

Mike: OK. For how long?

Chef: About 5 minutes.

Mike: And then?

Chef: Peel them and remove the seeds, we'll make grape juice today.

Mike: I see. It's done.

Chef: Put them in the liquidizer.

A: Mike, What are you doing?

B: I'm *peeling the peach.*

(make; juice/wash; cherry/ soak; strawberry/ remove; seeds)

Activity 4 Listen and complete.（听录音，完成对话）

A. Peel and cut up the fruits

B.We'll make a fruit salad

C. a cup of yogurt in the salad

D.What comes next?

Mike: What are we going to do today?

chef: ______________. First, prepare one orange, two bananas and an apple.

Mike: Oh, I see. And then?

Chef: ____________, then put the fruits in a mixing bowl.

Mike: Okay, I've finished. _____________?

Chef: Put two teaspoons of honey and___________.

Mike: Mix them up?

chef: That's right.

Activity 5 Pair work.（对话练习）

eg Mike is talking about fruits ...with Tom.

Mike

Tom

Part Two

Activity 1 Read.（读单词和词组）

kiwi fruit 猕猴桃	pineapple 菠萝	lychee 荔枝	papaya 木瓜
star fruit 杨桃	dragon fruit 火龙果	mango 芒果	coconut 椰子

Activity 2 Look, write and talk.（看图写英文并进行对话练习）

A : What fruit do you like to eat?

B : I like to eat __________.

 ____________

 ____________

 ____________

 ____________

 ____________

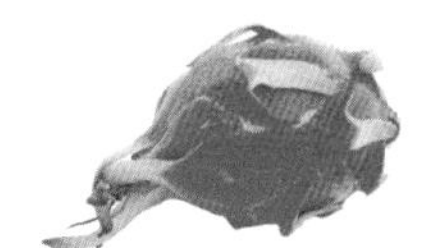 ____________

 ____________

 ____________

Activity 3 Complete and read.（完成并朗读对话）

A. wash and peel them	B. remove the seeds
C. make peach juice	D. put them in the liquidizer

Chef : We'll __________________ today.

Mike : That's wonderful. What shall we do?

Chef : First __________________________.

Then ___________________________.

After that ______________________.

Mike : OK. I understand.

Part Three

Activity 1 Tick.（勾出你喜欢吃的坚果）

walnut 核桃	peanut 花生	chestnut 栗子
hazelnut 榛子	almond 杏仁	cashew 腰果

Activity 2 Look and write.（看图写英文）

  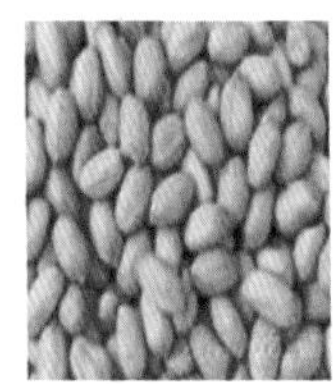   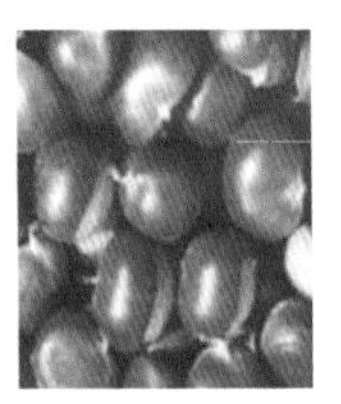

______ ______ ______ ______ ______ ______

Activity 3 Look at the following words.（辨识单词）

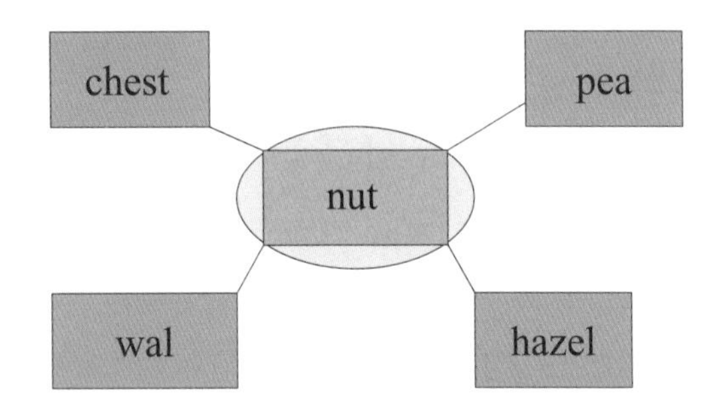

Are they easy to remember?

Activity 4 Fill in and evaluate.（填空并自我评价）

1. w_term_lon 2. p_ _ch 3. p_ _r 4. m_ngo 5. s_ _d
6. p_ne_pple 7. str_wb_rry 8. l_ch_e 9.p_p_ y_ 10. c_sh_w

Assessment : If you can write 8-10 words, you are perfect.
If you can write 4-7 words, you are good.
If you can write 1-3 words, you will try again.

Give it off（知识拓展）

Practice 1 Match.（连线）

A

B 

C 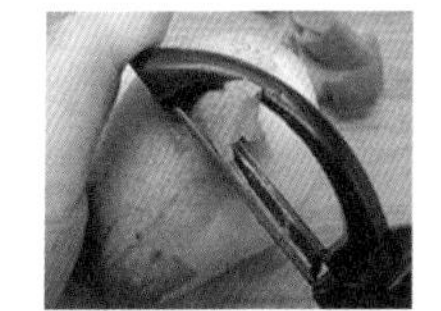

D 

E 

Practice 2 Summarize.（总结水果和坚果单词）

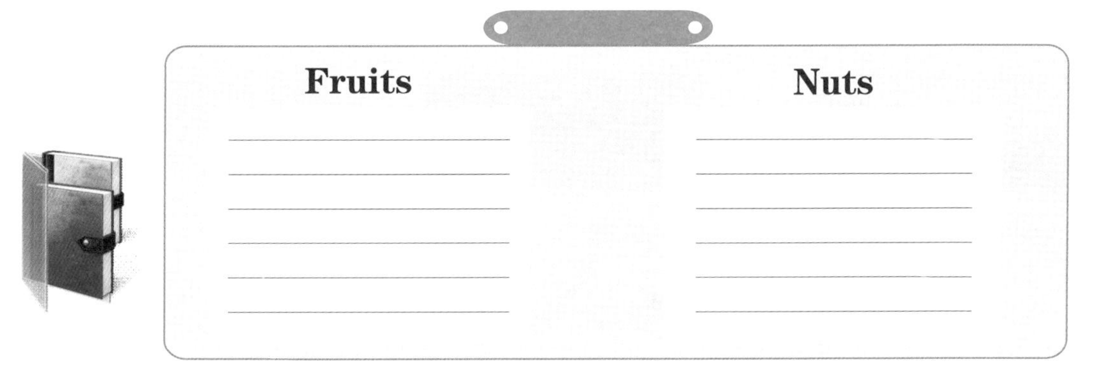

Practice 3 Make dialogues.（对话练习）

A: What are you doing?
B: I'm ________________.
A: For what?
B: ____________________.

A: What's the color of the grapes?
B: It's __________________.
A: Do you like them?
B: ____________________.

A: Do you like watermelon?
B: ____________________.
A: In which season do we have them?
B: ____________________.

A: What's this?
B: It's __________________.
A: Is it sweet or sour?
B: ____________________.

Practice 4 Write.（根据所给的颜色，写出对应的水果或坚果）

Unit exercises（单元练习）

Exercise 1 Describe and write.（根据所给的信息，用英文描述水果沙拉的制作过程）

Ingredients

First ______________________________.

Next ______________________________.

Then ______________________________.

After that ______________________________.

Finally ______________________________.

Exercise 2 Complete the expressions.（完成短语）

1. ____________ a fruit salad（做）
2. ____________ the grapes（浸泡）
3. ____________ the peaches（去皮）
4. ____________ the seeds from the oranges（去除）
5. ____________ the watermelon（榨汁）
6. ____________ the fruits into the colander（放）
7. ____________ the pineapple（切丁）
8. ____________ the mangoes（切成片）

Exercise 3 Tick.（勾出不同类别的单词或词组）

1. A. peach B. apple C. banana D. peanut
2. A. chestnut B. apricot C. grape D. star fruit
3. A. pineapple B. almond C. melon D. cherry
4. A. mango B. papaya C. strawberry D. kiwi fruit
5. A. watermelon B. pear C. grape D. tomato
6. A. orange B. strawberry C. peach D. potato
7. A. walnut B. pineapple C. hazelnut D. peanut

Exercise 4 Name the fruits.（根据所给信息，写出对应的水果名称）

1. Name a fruit that starts with an "A". It is white on the inside and can be red, yellow,or green on the outside. ( )
2. Name a long, thin fruit that starts with a "B". It is yellow on the outside and white on the inside. ( )
3. Name a sweet fruit that grows in bunches (一串) on vines (藤). It starts with a "G". ( )
4. Name a small round red fruit that starts with a "C". ( )

Exercise 5 Discuss and share.（讨论并和同学分享答案）

1. What fruits do you like best?

____.

2. In which season do we have watermelon?

____.

3. Do you like to eat mangoes?

____.

4. Which do you prefer, grapes or apples?

____.

5. How do you wash the strawberries?

____.

Learning tips（学习提示）

师傅说了，菜板是按颜色分类的，同时果蔬的消毒要严格按照下面的比例！

生肉类	海鲜	果蔬
奶酪	面包	熟食

84 消毒液比例		
原液：水	1：250	1：300
时间：	10 分钟	10 分钟
类别：	餐具	瓜果蔬菜

Culture knowledge（文化知识）

水果小知识

饭前还是饭后吃水果好?

水果属于低热量食物，水分含量高，并含有比蔬菜多的以双糖或单糖形式存在的碳水化合物。因此，饭前吃水果可以增加饱腹感，有利于控制食物的摄入总量，而若饭后马上吃水果，相当于在已基本吃饱的情况下再增加食物摄入量，易造成热量摄入超标，不利于控制体重。因此，对于节食的减重者来说，吃水果宜在饭前半小时。

一日一苹果，医生远离我

你可能知道这句熟悉的谚语：“一天一个苹果，医生远离我。” 虽然研究表明，多吃苹果与少生病没有实际的关系，但是在饮食中添加苹果，有利于身体的健康。苹果富含营养，能够促进心脏长期健康，还能降低患某些癌症的风险。零食，吃个苹果吧，不要吃薯条了！

Unit 7

Vegetables

Learning goals（学习目标）

You will be able to

★ get familiar with the English names of vegetables.

★ know the basic sentences for asking vegetables.

★ describe the usage of vegetables.

Word list（单词表）

Key words & expressions（重点单词和词组）

onion ['ʌnjən]	洋葱
celery ['seləri]	芹菜
Chinese cabbage	大白菜
eggplant ['egplɑ:nt]	茄子
lotus root ['ləʊtəs ru:t]	莲藕
cucumber ['kju:kʌmbə (r)]	黄瓜
taro ['tɛroʊ]	芋头
soybean ['sɔɪˌbi:n]	黄豆
chickpea ['tʃɪkpi:]	鹰嘴豆
broad bean [ˌbrɔ:d 'bi:n]	蚕豆
cowpea ['kaʊpi:]	豇豆
turnip ['tɜ:nɪp]	白萝卜
cauliflower ['kɒliflaʊə (r)]	菜花
red pepper [ˌred 'pepə (r)]	红椒
zucchini [zu'ki:ni]	西葫芦
yam [jæm]	山药
carrot ['kærət]	胡萝卜
lettuce ['letɪs]	生菜
spinach ['spɪnɪtʃ]	菠菜
parsley ['pɑ:sli]	欧芹
pumpkin ['pʌmpkɪn]	南瓜
sweet potato [ˌswi:t pə'teɪtəʊ]	红薯
green bean [bi:n]	青豆
red bean [red bi:n]	红豆
haricot ['hærɪkəʊ]	扁豆
pea [pi:]	豌豆
asparagus [ə'spærəgəs]	芦笋
broccoli ['brɒkəli]	西兰花
mushroom ['mʌʃrʊm]	蘑菇
artichoke ['ɑ:tɪtʃəʊk]	洋蓟
olive ['ɒlɪv]	橄榄

Expanded words & expressions（拓展单词和词组）

mixing bowl	拌菜碗	be fit [fɪt] for	适合于
finely	精细地	hygiene ['haɪdʒiːn]	卫生
grind [graɪnd]	磨碎；压碎	carve [kɑːv]	雕刻
pay attention to	注意	scrub [skrʌb]	刷洗；擦洗
rinse [rɪns]	冲洗；漂洗	completely [kəm'pliːtli]	彻底地；完全地
thoroughly ['θʌrəli]	非常；极其	the rest of	其余

Lead in（导入）

Activity 1 Look and match.（看图并连线）

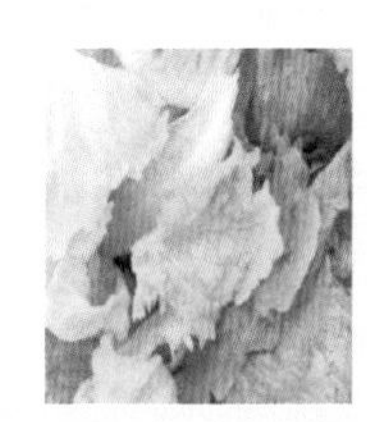

onion　　carrot　　celery　　lettuce　　Chinese cabbage

Activity 2 Write.（根据描述写出下列蔬菜的英文表达）

1.________ What is long and orange?

2.________ What is long, big and white inside, but green leaves outside?

3.________ Used in salad as they are crisp and fresh with big leaves.

4.________ What is purple and spicy?

Activity 3 Oral practice.（口语练习）

A: What is it called in English?
B: It is *celery.*
A: What do we call it in Chinese?
B: It's *"qincai".*

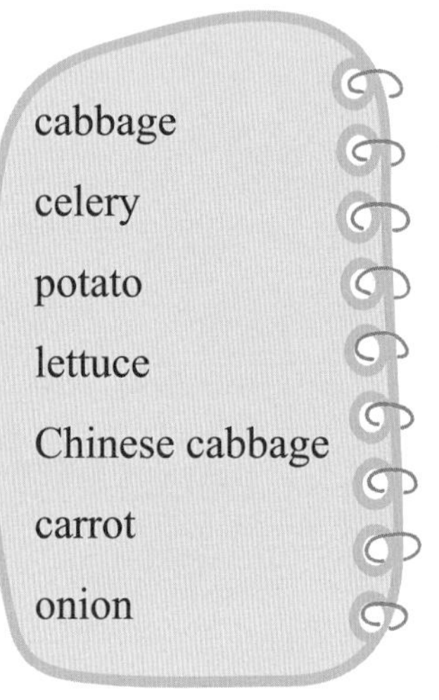

Focus on learning（聚焦学习）

Part One

Activity 1 Read.（读单词和词组）

spinach 菠菜	eggplant 茄子	parsley 欧芹	lotus root 莲藕
pumpkin 南瓜	cucumber 黄瓜	sweet potato 红薯	taro 芋头

Activity 2 Look and write.（看图写英文）

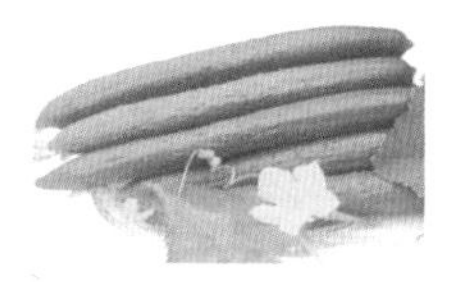

Activity 3 Read and answer.（读对话，回答问题）

Questions: 1. What does the chef want Mike to do?

2. Where will Mike put onions?

Mike : What can I do for you?

Chef : Give me some carrots and onions.

Mike : OK. What shall I do with them?

Chef : Slice the onions carefully.

Mike : And then what about after that?

Chef : Dice the carrots.

Mike : Shall I put them on the chopping board?

Chef : Well,you'd better put them in the mixing bowl.

Mike: OK. I see.

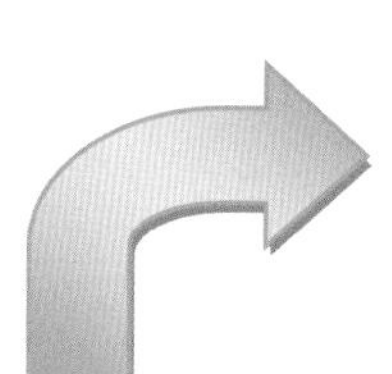

A : Mike, what will you do?

B : I'll ______ the ________.

(chop/cabbage; slice/celery; dice/potato; grind/ garlic)

Activity 4 Listen and complete.（听录音，完成对话）

A. slice them	B. dice the celery
C.cut onions and carrots	D. make vegetable soup

Mike :What are we going to do?
Chef : We are going to ______________.
Mike : Shall I ____________?
Chef : Yes, ________________finely.
Mike : And then ____________?
Chef : Yes, dice the celery and chop the mushroom on the chopping board.
Mike : I see.

Activity 5 Pair work.（对话练习）

eg Mike is talking about some vegetables with Tom.

Mike

Tom

Part Two

Activity 1 Read.（读单词和词组）

green bean 青豆	soybean 黄豆	red bean 红豆	chickpea 鹰嘴豆
haricot 扁豆	broad bean 蚕豆	pea 豌豆	cowpea 豇豆

Activity 2 Look, write and talk.（看图写英文，并进行对话练习）

eg

A : Which bean can be used for making soup?

B: ___________.

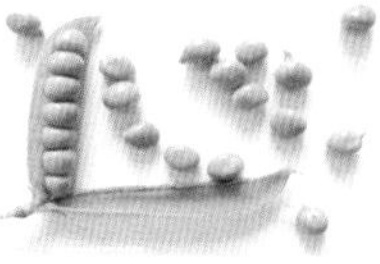

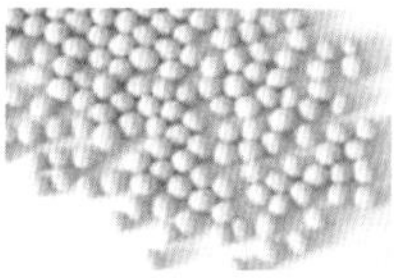

___________ ___________ ___________ ___________

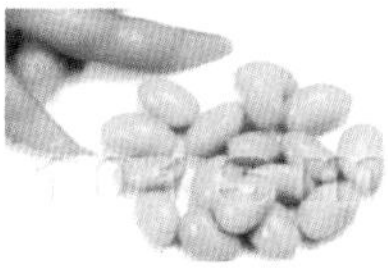

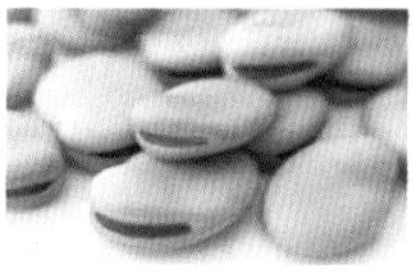

___________ ___________ ___________ ___________

Activity 3 Complete and read.（完成并朗读对话）

A. put them in the mixing bowl

B. Slice them finely

C. Peel them carefully

D. wash four eggplants

Mike : What are we going to do?

Chef : First, ___________________________________.

Mike : What should I do next?

Chef : ___________________________________.

Mike : Then what about after that?

Chef : ___________________________________.

Mike : OK. I've finished.

Chef : Finally, ___________________________________.

Part Three

Activity 1 Tick.（勾出你晚餐经常食用的蔬菜）

asparagus 芦笋	turnip 白萝卜	broccoli 西兰花	
cauliflower 菜花	mushroom 蘑菇	red pepper 红椒	
artichoke 洋蓟	zucchini 西葫芦	olive 橄榄	yam 山药

Activity 2 Look and write.（看图写英文）

________ ________ ________ ________ ________

________ ________ ________ ________ ________

Activity 3 Fill in and evaluate.（填空并自我评价）

1. l_m_n	2. p_mpk_n	3. l_tt_ce	4. br_ccol_	5. c_bb_ge
6. m_shr_ _m	7. c_c_mb_r	8. gr_nd	9. d_c_	10. c_l_ry
11. t_r_	12. ca_lifl_wer	13. p_rsl_y	14. _ggpl_nt	15. y_m

Assessment : If you can write 10-15 words, you are perfect.

If you can write 7-9 words, you are good.

If you can write 1-6 words, you will try again.

Give it off（知识拓展）

Practice 1 Match.（连线）

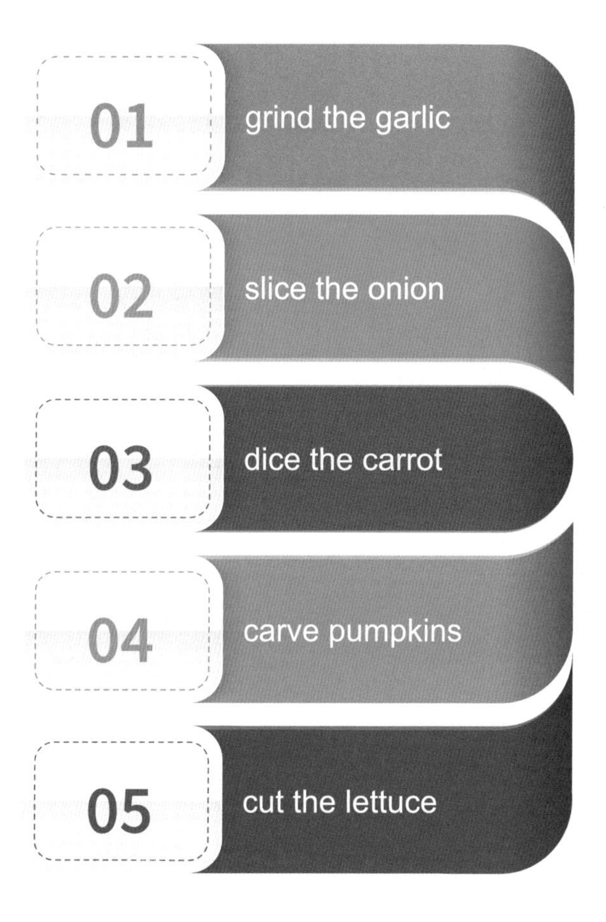

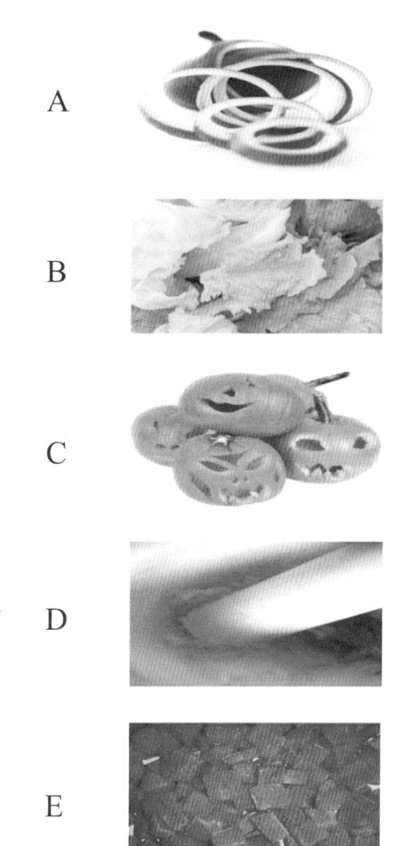

Practice 2 Summarize.（总结本单元的重点句型）

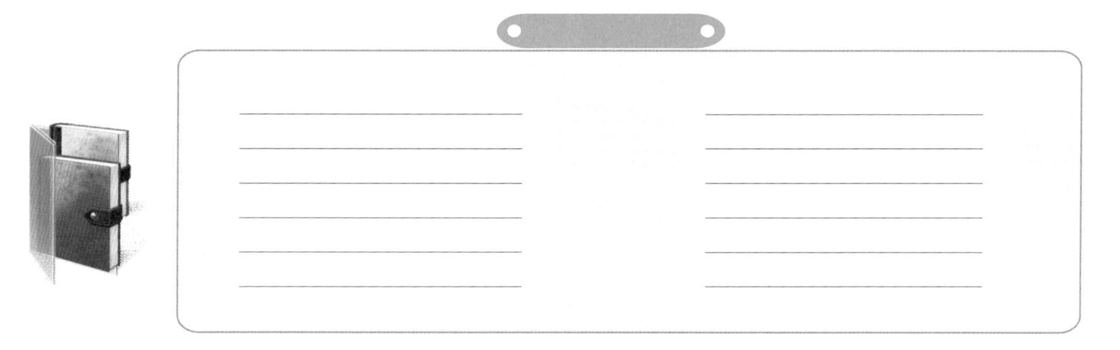

Practice 3 Make dialogues.（对话练习）

A:What will you do?

B: *Chop the celery.*

A: For what?

B: *To make a salad.*

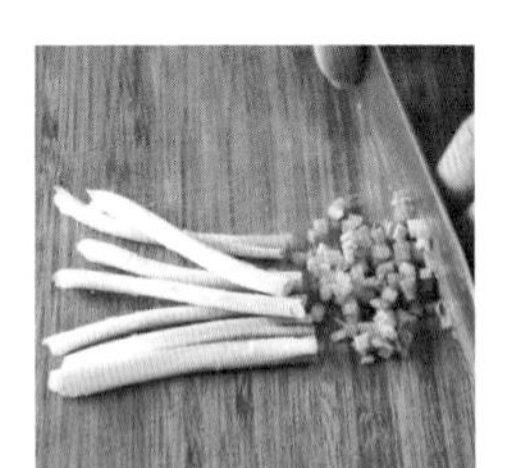

Practice 4 Fill in the blanks.（填空）

add	pour	grind	slice	put...on

1. ______ away the dirty water.
2. The fire is going out. Will you ______ some wood?
3. Please ______ some sugar ______ the broccoli.
4. Would you ______ some garlic finely for the stew?
5. —Shall I chop the carrots?
   —No, ______ carefully.

Unit exercises（单元练习）

Exercise 1 Match.（连线）

remove	the carrots down the middle
dice	the ends
chop	the seeds from the pumpkin
grind	the sweet potatoes carefully
peel	the cucumbers
slice	the parsley finely
scrub	the onions finely
cut ... into two	the garlic
trim off	the zucchini
split	the red peppers

Exercise 2 Make sentences in right order.（连词成句）

1. is, the, most, salads, of, lettuce, "heart", of

____.

2. slicing, pumpkins, are, now, you

____?

3. I, shall, dice, yams, the

____?

4. for, chop, onions, the, salad, the, please

____.

5. cook, I, shall, how, artichoke, the

____?

Exercise 3 Fill in the blanks.（根据所给信息填空）

First wash some  _______ and _______, then chop them carefully with some _______ on the _______. And then put them together with _______ and _______ in a _______. Add some salt, mayonnaise, and some _______ on them. Now you have a delicious _______ salad. Do you like it?

Exercise 4 Translate into English.（将下列句子翻译成英文）

1. 我们做汤需要切得很细的洋葱。

______________________________.

2. 我现在切芹菜丁好吗？

______________________________?

3. 他正在捣蒜泥。

______________________________.

4. 请先将胡萝卜洗净，然后切成丝。

______________________________.

5. 你知道如何烹饪洋蓟吗？

______________________________?

Exercise 5 Classify and write.（将蔬菜分类，并写在横线上）

soybean	lettuce	sweet potato	zucchini	chickpea
celery	cowpea	spring onion	turnip	Chinese cabbage
spinach	yam	asparagus	carrot	parsley
taro	pumpkin	lotus root	pea	bean
red bean	green bean	broad bean		

根茎类蔬菜 ______________________________

叶类蔬菜 ______________________________

豆类蔬菜 ______________________________

Learning tips（学习提示）

Hygiene is very important, so you must pay attention to cleaning hands and vegetables before cooking.

How to wash hands

1. Wet hands with WARM water.
2. Soap and scrub for 20 seconds.
3. Rinse under clean, running water.
4. Dry completely using a clean cloth or paper towel.

How to clean vegetables

1. Gently rub vegetables under running water.
2. Remove the outer leaves of lettuce and cabbage heads, and thoroughly rinse the rest of the leaves.

Culture knowledge.（文化知识）

蔬菜的营养素与功效

蔬菜含有丰富的碱性矿物元素与膳食纤维等。丰富抗氧化的蔬菜可以抗癌，特殊香味的蔬菜可以降血压，含有纤维素的蔬菜可以促进肠胃的正常蠕动，降低体内胆固醇水平，调节糖尿病患者的血糖水平等功效。应当尽量避免没有蒸熟，以避免生肉中含有的致病菌危害健康。

烹饪方法影响蔬菜的营养成分

在烹饪过程中，应该将不同品种蔬菜的原料与所含的营养成分进行比较，再将食物合理搭配，采用科学方法烹饪，尽量减少蔬菜在烹饪中营养素的流失。

Unit 8

Meat

Learning goals（学习目标）

You will be able to

★ get familiar with different kinds of meat in the kitchen.

★ know how to ask the ways of cooking.

★ describe four special parts of beef.

Word list（单词表）

Key words & expressions（重点单词和词组）

English	中文	English	中文
pork [pɔ:k]	猪肉	chicken ['tʃɪkɪn]	鸡肉
beef [bi:f]	牛肉	mutton ['mʌtn]	羊肉
T-bone steak	T 字骨牛排	sirloin ['sɜ:lɔɪn]	西冷（外脊）牛肉
tenderloin ['tendələɪn]	里脊肉	rib [rɪb] eye steak	肋眼牛肉
goose [gu:s]	鹅	turkey ['tɜ:ki]	火鸡
pigeon ['pɪdʒɪn]	鸽子	quail [kweɪl]	鹌鹑
guinea fowl ['gɪni faʊl]	珍珠鸡	deer [dɪə (r)]	鹿
boar [bɔ:(r)]	野猪	pheasant ['feznt]	野鸡 / 雉
hare [heə (r)]	野兔	veal kidney [vi:l 'kɪdni]	小牛肉肾
Osso Bucco ['ɒsəʊ' bʊkəʊ]	小牛骨 / 红烩牛膝	oxtail ['ɒksteɪl]	牛尾
foie gras [ˌfwɑ: 'grɑ:]	肥鹅肝		

Expansion words & expressions（拓展单词和词组）

cattle ['kætl]	牛	tough [tʌf]	坚硬的
tender ['tendə (r)]	嫩的	sauté ['səʊteɪ]	炒
necessary ['nesəsəri]	必要的	mean [miːn]	意思是
typical ['tɪpɪkl]	典型的	invite [ɪn'vaɪt]	邀请
liver ['lɪvə (r)]	（动物供食用的）肝	toast [təʊst]	吐司

Lead in（导入）

Activity 1 Look and match.（看图并连线）

pork　　chicken　　beef　　mutton

Activity 2 Write.（参考 Activity 1，根据描述写出英文表达）

1. ____________ We get it from the pig.
2. ____________ We get it from the sheep or goat.
3. ____________ We get it from the cock or hen.
4. ____________ We get it from the cattle.

Activity 3 Oral practice.（口语练习）

A: Do you like to eat __________?
B: Yes, I do. (No, I don't.)

pork
chicken
beef
mutton

Focus on learning（聚焦学习）

Part One

Activity 1 Read.（读单词和词组）

T-bone steak T 字骨牛排
sirloin 西冷（外脊）牛肉
tenderloin 里脊肉
rib eye steak 肋眼牛肉

Activity 2 Look and write.（看图写英文）

__________ __________ __________ __________

Activity 3 Read and answer.（读对话，回答问题）

Questions: 1. Which part is a little tough among the four parts?

2. Is it more delicious to grill tenderloin? Why?

Mike: What shall we do with the tenderloin?

Chef: It's the most tender part of the beef. We usually sauté or grill it.

Mike: How about the rib eye steak or T-bone?

Chef: It's more tender. It is more delicious to roast or grill it.

Mike: Is the sirloin tender too?

Chef: It's a little tough, but young people think the sirloin steak tasty.

A: Mike, how to say "里脊肉 / 肋眼牛肉" in English?

B: We call it ___________.

A: Is it the most tender part?

B: Yes, that's right./No, I don't think so.

Activity 4 Listen and complete.（听录音，完成对话）

A. Let's grill it	B. Why not cook for a long time
C. The meat is very tender	D. roast for a short time

Mike : What shall I do with the tenderloin?

Chef : ______________.

Mike: What about sauting it?

Chef : That's good. And you can also________.

Mike : ___________________________?

Chef : ___________.It's not necessary to do so.

Mike : I see.

Activity 5 Pair work.（对话练习）

eg Mike is talking about different parts of beef with Tom.

Mike

Tom

Part Two

Activity 1 Read.（读单词和词组）

goose 鹅	turkey 火鸡	pigeon 鸽子	quail 鹌鹑	guinea fowl 珍珠鸡
deer 鹿	boar 野猪	pheasant 野鸡 / 雉	duck 鸭子	hare 野兔

Activity 2 Look, write and talk.（看图写英文，并进行对话练习）

A : What can you see in the picture?　B : I can see a/some ______.
A : What do you mean by "_____" in Chinese?　B : It means ______.

__________ __________ __________ __________ __________

__________ __________ __________ __________ __________

Activity 3 Complete and read.（完成并朗读对话）

A. What are we going to have	B. Let's taste the typical Beijing food
C. invite you to have a dinner	D. Beijing Roast Duck

Chef: I'd like to ________________.
Mike: Wonderful! __________?
Chef: ________________________.
Mike: What is the special food?
Chef: ________________.
Mike: How nice! Let's go now.
Chef: Er..., but there is a problem. I have no money.

Part Three

Activity 1 Tick.（勾出你喜欢吃的食物）

veal kidney 小牛肉肾

Osso Bucco 小牛骨 / 红烩牛膝

oxtail 牛尾

foie gras 肥鹅肝

Activity 2 Look and write.（看图写英文）

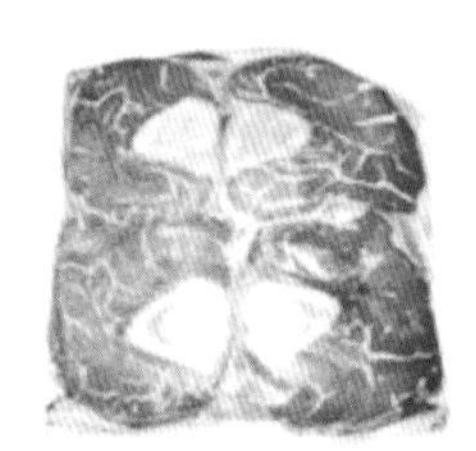

______ ______ ______ ______

Activity 3 Decide true or false.（判断正误）

(   ) 1. Veal kidney is an inner part of the young cattle.

(   ) 2. Osso Bucco is a kind of plant.

(   ) 3. Oxtail is the tail of the cattle.

(   ) 4. Foie gras is the liver of the goose.

Activity 4 Fill in and evaluate.（填空并自我评价）

1. b_ _f　　2. m_ tt _n　　3. ch _ck _n　　4. d_ck　　5. g_ _se

6. t_ _key　　7. p_ge_n　　8. qu_ _l　　9. d_ _r　　10. p_ _k

Assessment : If you can write 8-10 words, you are perfect.

If you can write 4-7 words, you are good.

If you can write 1-3 words, you will try again.

Give it off（知识拓展）

Practice 1 Match.（连线）

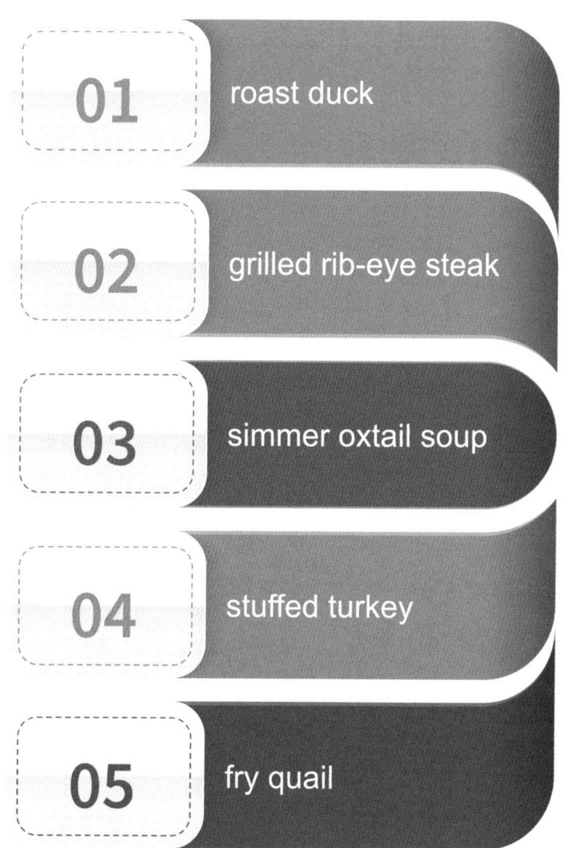

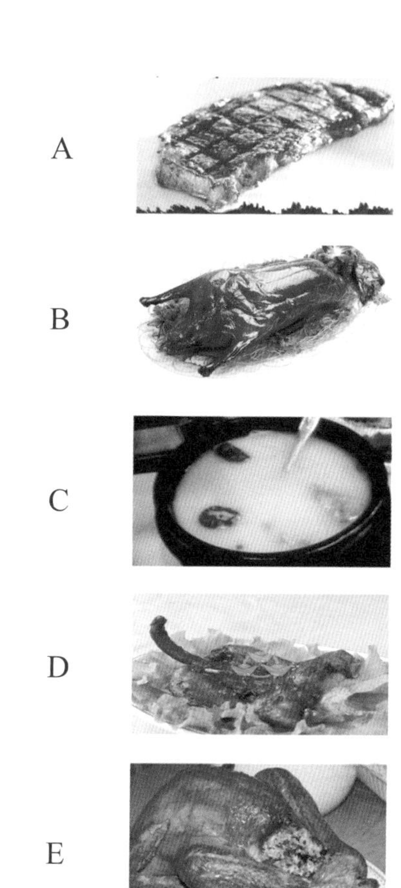

Practice 2 Summarize.（总结本单元的重点句型）

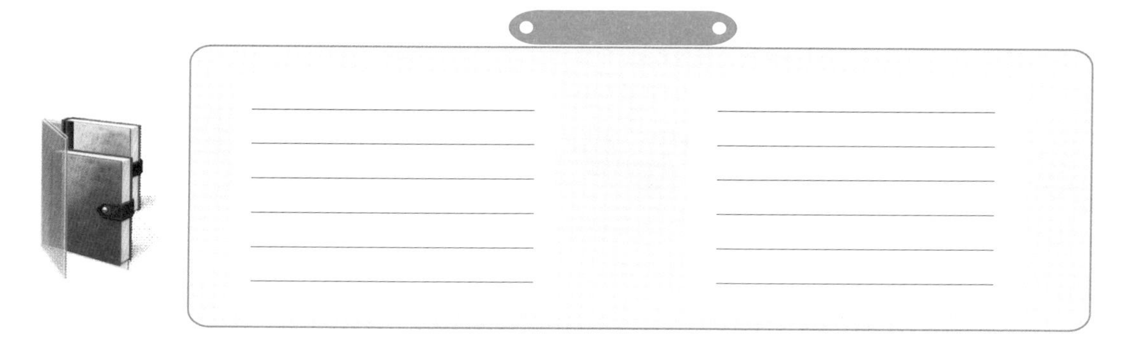

Practice 3 Make dialogues.（对话练习）

A: How do you call this?
B: It is ____________________.
A: What shall I do with it?
B: It is better to ____________________.

A: What do you call this?
B: It is _______________.
A: How shall I do with it?
B: You'd better _________.

A : How do you call this in English?
B: It is _______________.
A: What is the best way to cook it?
B: I think you can _________.

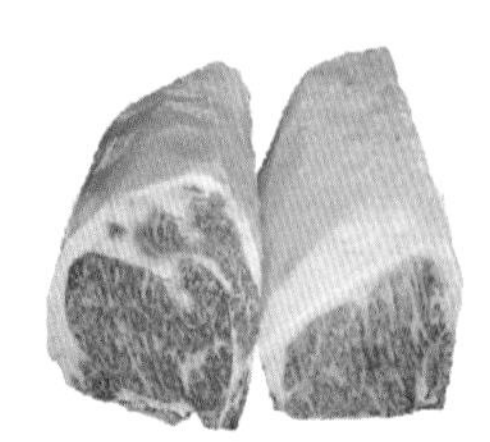

A: What do you mean by sirloin in Chinese?
B: It is called_______________.
A: Do you know how to cook it?
B: I can ___________.

Practice 4 Fill in the blanks.（选词填空）

fry 煎	braise 焖	grill 扒	stew 烩	sauté 炒	roast 烤

1. I'm similar to "fly", but my second letter is "r". ____________.
2. "Toast" is a kind of bread. Please change the first letter with "r". ___________.
3. You can find my first name in "steak", my last name is in "new", ________.
4. Don't take "gr" away, I'll be "ill" right away. ____________.
5. My meaning is to cook for a long time over the low fire. ____________.
6. I'm from France, meaning to cook for a short time. ____________.

Unit exercises（单元练习）

Exercise 1 Choose the right answers.（选择正确答案）

1. Which one is not meat?________.

   A. Pork　　B. Beef　　C. Braise　　D. Chicken

2. When you cook tough meat, you'd better not ______ it.

   A. stew　　B. braise　　C. sauté　　D. roast

3. _______ is a kind of birds.

   A. Deer　　B. Quail　　C. Boar　　D. Oxtail

4. The inner part of the cattle is _________.

   A. veal kidney　　B. Osso Bucco　　C. tenderloin　　D. rib eye steak

5. Western cooking uses _______ the most.

   A. duck　　B. beef　　C. quail　　D. mutton

Exercise 2 Complete the following sentences.（完成句子）

1. What shall I do with the beef?

_____ shall I ____ with the chicken?

2. Silver side is tender, loin is more tender, fillet is the most tender.

Apple is delicious, cherry is ____ _____, strawberry is _____ _____ _____.

3. Why not cook the meat for a long time?

_____ _____ ask the executive secretary for advice?

4. It's not necessary to freeze the mutton. I'll fry it.

_____ ______ _____ ______ peel the ginger. ____ stew the meat with them.

5. I'd like to invite you to have a dinner.

_____ ______ _____ invite you ____ visit our hotel.

6. Let's try typical Beijing food.

_____ ______ typical western food.

Exercise 3 Tick the different words.（勾出不同类型的单词或词组）

( ) 1. A. beef B. pork C. chicken D. mutton

( ) 2. A. goose B. duck C. turkey D. tenderloin

( ) 3. A. oxtail B. turkey C. foie gras D. veal kidney

( ) 4. A. deer B. rib eye steak C. sirloin D. beef T-bone

( ) 5. A. boar B. hare C. pheasant D. quail

( ) 6. A. beef steak B. roast C. grill D. pan fry

Exercise 4 Translate into Chinese or English.（句子中英文翻译）

1. 我最喜欢吃牛肉。

____

2. 这鸭肉有点儿老，用焖或炖的方法比较好。

____

3. 我们为什么不自己试着做牛排呢?

____

4. It's not necessary to grill the meat for such a long time.

____

5. I'd like to invite you to try some French mashed foie gras.

____

Learning tips（学习提示）

Name the signs which are about food quality or safety.

______ ______ ______ ______ ______ ______ ______ ______

A. 质量安全标志	B. 无公害农产品标志
C. 保健食品标志	D. 国家免疫产品标志
E. 绿色食品标志	F. 食品安全标志
G. 有机食品标志	H. 有机（生态）产品标志

Culture knowledge（文化知识）

1. 锅具的选择

在餐厅吃牛排时，我们经常会看到端上来的牛排上有着一条条的烧烤花纹，这其实是煎烤时所选用的锅具是否带有纹路导致的。在煎牛排时，肉汁会流出来，所以条纹煎锅里凸起的条纹就是让多余的汁水流到凹处，从而保持牛排的温度，不至于被烤糊。

用平底锅煎出的牛排虽然没有漂亮的花纹，但是由于它的接触面积比较多，受热面积更大，因此更容易煎出均匀的焦化外层，锁住所谓的肉汁，达到外焦里嫩的效果。

2. 温度与时间的控制

在煎牛排前，锅一定要烧得非常热，这样牛排在下锅的瞬间，高温会迅速把肉的表面封住，保证肉里的汁水充盈。牛排煎制的时长，取决于牛排的厚度及想要的熟度。

3. 封汁静置的好处

牛排煎完后，可以在锅中再加入 3 根百里香和一块黄油。倾斜锅子，先单独煸香黄油！然后用勺子不停地把黄油淋在牛排的表面，让牛肉吸饱香味和热量，翻面再煎 45 秒。煎好的牛排可以盖上锡纸或者盘子，小块牛排室温下静置 5 到 10 分钟，大块静置 10 到 15 分钟。这一步静置可以让煎出的牛排纤维松弛，因而牢牢地锁住肉汁。

Unit 9

Seafood

Learning goals（学习目标）

You will be able to

★ get familiar with different seafood.

★ know the preparation of seafood in English.

★ tell the process of making seafood dish.

Word list（单词表）

Key words & expressions（重点单词和词组）

trout [traʊt]	鳟鱼	 white bait [beɪt]	银鱼 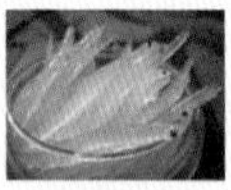
sturgeon ['stɜːdʒən]	鲟鱼	 salmon ['sæmən]	三文鱼 
mandarin ['mændərɪn] fish	鳜鱼	 flat fish	比目鱼 
cod fish	鳕鱼	 perch [pɜːtʃ]	鲈鱼 
anchovy ['æntʃəvɪ]	鳀鱼	 tuna ['tjuːnə]	金枪鱼 
grouper ['gruːpə]	石斑鱼	 sardine ['sɑːdiːn]	沙丁鱼
caviar ['kæviɑː(r)]	鱼子酱	 prawn [prɔːn]	对虾 
lobster ['lɒbstə]	龙虾	 oyster ['ɔɪstə]	牡蛎 
crab [kræb]	螃蟹	 scallop ['skɒləp]	扇贝 
mussel ['mʌsl]	贻贝	 abalone [ˌæbə'ləʊnɪ]	鲍鱼 
conch [kɔŋk]	海螺	 bream [briːm]	鳊鱼 
mackerel ['mækərəl]	鲭鱼	 eel [iːl]	鳝鱼 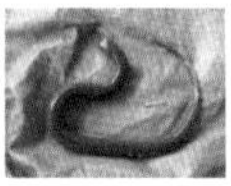
herring ['herɪŋ]	鲱鱼 		

Expanded words & expressions（拓展单词和词组）

fish bone [bəʊn]	鱼骨头	fish forceps ['fɔːsɪps]	鱼镊子
fish scaler ['skeɪlə]	刮鳞器	gut [gʌt]	内脏，取出内脏
take out	取出	special ['speʃəl]	特别的，特殊的
stomach ['stʌmək]	胃部，肚子	cut open	打开
de-head	去头	de-vein [veɪn]	去泥肠
split [splɪt]	劈开，撕开，切开	bone the fish	去鱼骨
marinade [ˌmærɪ'neɪd]	用腌泡汁浸泡的鱼	halibut ['hælɪbət]	大比目鱼

Lead in（导入）

Activity 1 Look and match.（看图并连线）

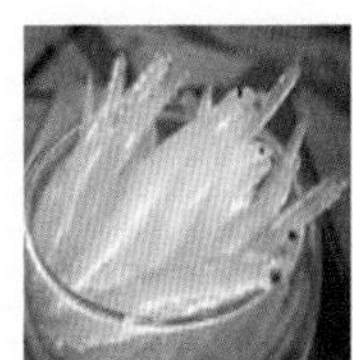

trout　white bait　sturgeon　salmon　mandarin fish

Activity 2 Decide true or false.（判断正误）

(　　) 1.Chinese sturgeon is very big and delicious.

(　　) 2.White bait is white with many fish bones.

(　　) 3. Mandarin fish often eats grass in river.

(　　) 4. Japanese like eating salmon.

(　　) 5. Red trout lives in the sea.

Activity 3 Oral practice.（口语练习）

A: What did you eat for dinner?

B: I ate ________.

salmon
trout
white bait
mandarin fish
sturgeon

Focus on learning（聚焦学习）

Part One

Activity 1 Read.（读单词和词组）

flat fish 比目鱼	cod fish 鳕鱼	perch 鲈鱼	anchovy 鳀鱼
tuna 金枪鱼	grouper 石斑鱼	sardine 沙丁鱼	caviar 鱼子酱

Activity 2 Look and write.（看图写英文）

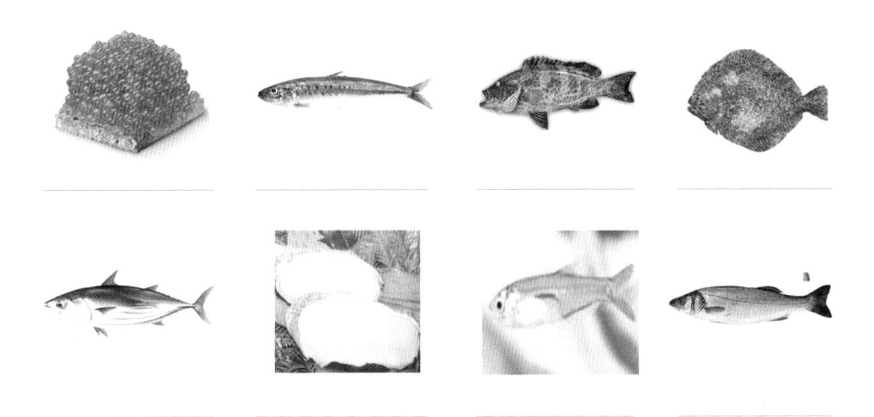

Activity 3 Read and answer.（读对话，回答问题）

Questions: 1. What can you see in the pictures?
2. What are they used for?

Mike : What are these?

Chef : They are fish forceps, fish scaler and fish scissors.

Mike : What shall I do with them?

Chef : You can scale the fish with the scaler, gut the fish with the scissors and take out the fish bones with the fish forceps.

Mike : They are very special.

Chef : Yes, you are right.

A: Mike,What shall I do with this (these)____?

B: You can _____________ with it (them).

(fish scissors/gut the stomach of the fish; fish scaler/scale the fish; fish forceps/take out the fish bone)

Activity 4 Listen and complete.（听录音，完成对话）

A.use these fish scissors	B.Scale it with your knife
C. cut open the stomach of the fish	D.take out the guts

Mike : What shall I do with the fish?

Chef : ________________, and then gut it.

Mike: Can I _____________?

Chef : Yes, they are very special. Now ___________ like that, then _________.

Mike : OK, I see.

Activity 5 Pair work.（对话练习）

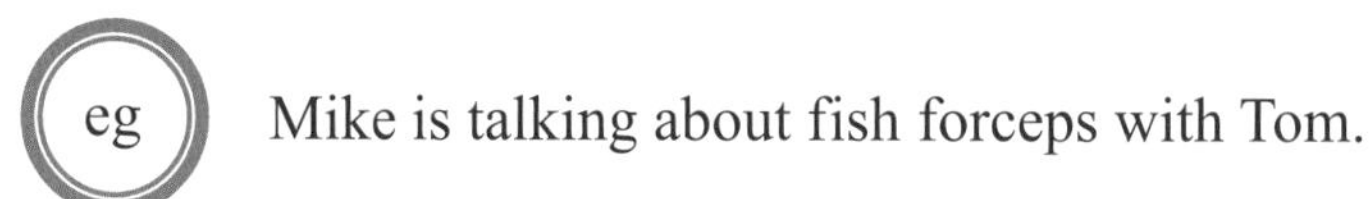

eg Mike is talking about fish forceps with Tom.

Mike

Tom

Part Two

Activity 1 Read.（读单词）

prawn 大虾，对虾	lobster 龙虾	oyster 牡蛎	crab 螃蟹
scallop 扇贝	mussel 贻贝	abalone 鲍鱼	conch 海螺

Activity 2 Look, write and talk in pairs.（看图写英文，并进行对话练习）

A: What shellfish would you like to eat?
B: I'd like to eat________.

____________ ____________ ____________ ____________

   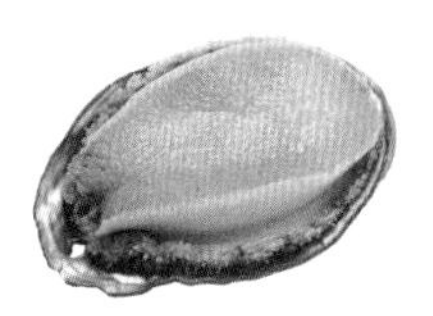

____________ ____________ ____________ ____________

Activity 3 Complete and read.（完成并朗读对话）

A. Let's de-head and de-vein	B. you should master the skills
C. use the oyster knife	D. Split it and bone the fish

Mike : Could you tell me how to treat shrimp?
Chef : ________________________.
Mike : How about the oyster?
Chef : You can __________________.
Mike : What about the fish?
Chef : __________________________.
Mike : OK, it is not difficult.
Chef : Yes,_________________________.

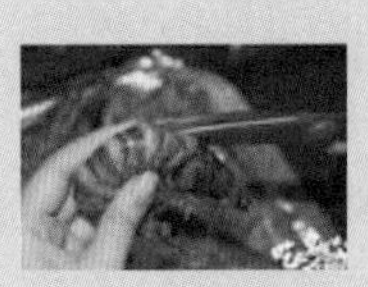

Part Three

Activity 1 Tick.（勾出你喜欢吃的鱼产品）

bream 鳊鱼　　mackerel 鲭鱼　　eel 鳝鱼　　herring 鲱鱼

Activity 2 Look and write.（看图写英文）

  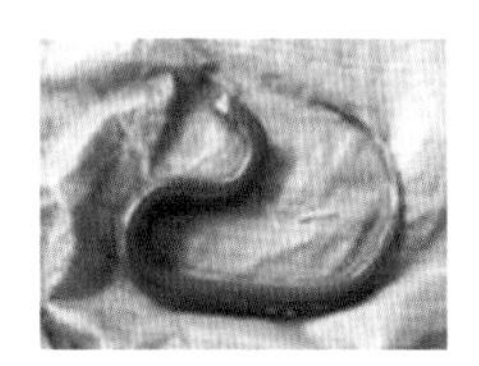 

________　________　________　________

Activity 3 Decide true or false.（判断正误）

(　　) 1. Never turn over the pieces of fish in the marinade.
(　　) 2. Pat the fish dry with towel.
(　　) 3. Scale fillet and bone the halibut with the fish scaler.
(　　) 4. Open the crab with the oyster knife.

Activity 4 Fill in and evaluate.（填空并自我评价）

1. _almon　2. _ardine　3. _turgeon　4. c_vi_r　5. fl_t fish
6. _nchovy　7. scall_p　8. spl_t　9. tun_　10. group_r

Assessment : If you can write 8-10 words, you are perfect.
If you can write 4-7 words, you are good.
If you can write 1-3 words, you will try again.

Give it off (知识拓展)

Practice 1 Match. (连线)

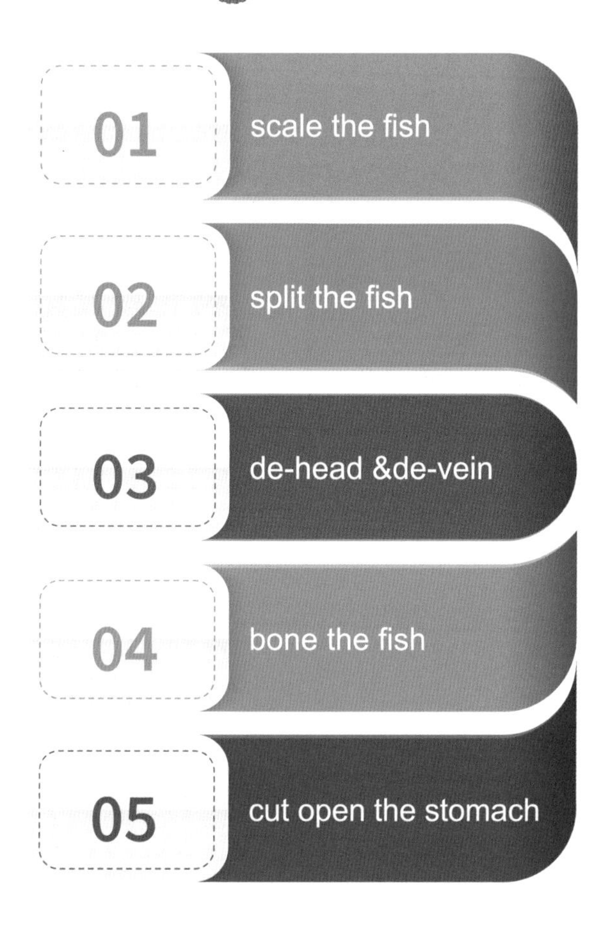

A 

B 

C 

D 

E 

Practice 2 Summarize. (总结本单元的重点句型)

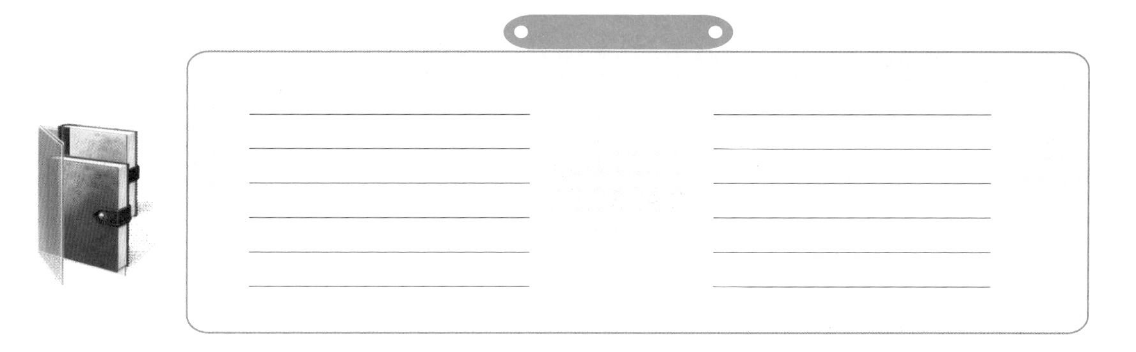

Practice 3 Make dialogues.（对话练习）

A:What shall I do with the fish?

B: *Scale it*.

A: With what?

B: *With a fish scaler*.

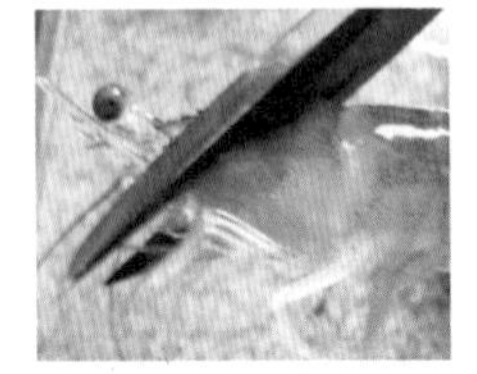

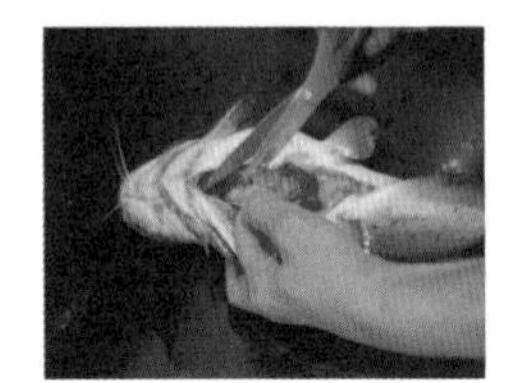

A: What have you cooked for this dinner?

B: We've cooked __________.

baked fish

curry mussel

fried shrimp

grilled tuna

oyster soup

scallop salad

Unit exercises（单元练习）

Exercise 1 Complete the following sentences.（完成句子）

1. We use fish scissors to ________________________.
2. We use a fish scaler to ________________________.
3. We use fish forceps to ________________________.
4. We use salmon knife to________________________.
5. We use oyster knife to ________________________.

Exercise 2 Tick the different words.(找不同类型的单词或词组)

1. A. fish scissors	B. fish scaler	C. fish forceps	D. salmon knife
2. A. crab	B. conch	C. mussel	D. trout
3. A. nip off	B. pull out	C. scale	D. eel
4. A. delicious	B. terrible	C. gut	D. wonderful
5. A. white bait	B. scallop	C. sardine	D. tuna

Exercise 3 Find the right answers.（找出正确答案）

(　　) 1. Could you tell me how to treat the shrimps before cooking?
(　　) 2. What shall we start with?
(　　) 3. What is the best way to clean it?
(　　) 4. How long shall I bake it?
(　　) 5. What is it served with?

A. The baked fish is served well with rice.
B. First you should scale the fish.
C. Peel off the shrimp sheels, nip off the head and pull out the vein.
D. Wash it in cold water.
E. For about 45 minutes.

Exercise 4 Translate into English.（将下列句子翻译成英文）

1. 用刮鳞器刮鱼鳞。

____________________________________.

2. 请用这把特别的剪刀剪开鱼腹。

____________________________________.

3. 我正在做牡蛎汤。

____________________________________.

Exercise 5 Write.（根据所给信息写出制作咖喱贻贝所需的原料、调料和制作过程）

Ingredients

__________ __________ __________ __________

Condiments

__________ __________ __________ __________

Process

Melt ____（黄油），add _____（洋葱）and cook until brown.

（加入面粉）_______，curry water to ________（混合）smoothly.

Add_____（贻贝）and make sure they're just covered by ____（水）.

Season with ________（盐）and_________（胡椒）.

Cook slowly at 325 degrees for 30_________（分钟）.

Learning tips（学习提示）

We'd better freeze seafood in the refrigerator or freezer. Let's look at the rule.

The TWO-hour Rule

Refrigerate perishable foods so TOTAL time at room temperature is less than TWO hours or only ONE hour when temperature is above 90 degrees.

Culture knowledge（文化知识）

海鲜小知识

贝类：吃鲜的但别贪生

食用海鲜类美食，一定要做熟再吃，千万不要贪图新鲜而食用生海鲜。贝壳类食材容易感染诺罗病毒和霍乱弧菌。因此，在吃扇贝的时候，应当蒸熟，以避免生肉中含有的致病菌危害健康。

扇贝：新鲜扇贝壳有弹性

扇贝的味道很鲜美，营养很丰富，是海味中的三大珍品之一。将扇贝中白色的内敛肌晒干做成干贝也着实不错。买新鲜扇贝的时候一定要闻一闻，有坏臭的味道就很不新鲜了；要是有汽油或者煤油的味道，要小心，可能是受到甲基汞的污染；还要看一看，新鲜的扇贝壳色泽光亮有弹性。

虾：虾头就别吃了

虾是老少皆宜的美味食品，尤其是鲜活的虾经过烹调之后更是滋味甜鲜、口感脆嫩。虾虽然鲜美，可也要注意，虾的头部胆固醇含量较高，也容易残留一些重金属等，在虾大量繁殖的季节，虾籽也含有一部分胆固醇，食用时需要注意，虾头和虾籽尽量少吃。

附 录

参考答案

Unit 1 Title Used in the Kitchen & Floor Plan

Lead in（导入）

Activity 2 P1.cook P2.manager P3.trainee P4. kitchen helper

Focus on learning（聚焦学习）

Part One

Activity 2 executive chef 行政总厨师长
executive sous-chef 行政副总厨师长
coffee shop kitchen sous-chef 咖啡厅副厨师长
commissary sous-chef 备餐厨房副厨师长
banquet kitchen sous-chef 宴会厨房副厨师长
pastry & bakery sous-chef 西点厨房副厨师长
Italian kitchen sous-chef 意大利厨房副厨师长
chief steward 管事部经理

Activity 3 P1. executive chef P2. executive sous-chef
P3. banquet kitchen sous-chef P4. Italian kitchen sous-chef
P5. coffee shop kitchen sous-chef P6. commissary sous-chef
P7. pastry & bakery sous-chef P8. chief steward

Activity 4 Q1.He is a pastry & bakery sous-chef.
Q2.He is in charge of making pastries.

Activity 5 D B A C

Part Two

Activity 2 冷菜厨房 热菜厨房 肉食厨房 备餐厨房 西点厨房
coffee shop kitchen Italian kitchen Chinese kitchen

Activity 3 D C B A

Activity 4 1.T 2. F 3. T 4. F 5. T 6. T

Part Three

Activity 2 jacket,apron, black or plaid pants, chef hat, tie,towel, shoes

Activity 3 1.e 2.e 3.e 4.a 5.o 6.a 7.a 8.o 9.a 10.e

Give it off（知识拓展）

Practice 1 1.E 2.A 3.D 4.C 5.B

Practice 2 1.What does he do in the kitchen?

He is a pastry & bakery sous-chef.

2.What is he in charge of?

He is in charge of making pastries.

3.What do you think of your job?

I find it attractive.

4.What're are they doing now?

They're working for the Italian kitchen.

Practice 3 1.coffee shop kitchen sous-chef 咖啡厅副厨师长

2. chef de partie 厨师主管

3. demi chef 领班 4. commis chef 厨师

5. kitchen helper 帮厨 6.trainee 学员

Practice 4 1. steward 2. kitchen helper 3. butcher 4. chef de partie

Unit exercises（单元练习）

Exercise 3 1. He is in charge of pastry & bakery kitchen.

2. The coffee shop kitchen sous-chef is in the coffee garden.

3. The commis chef is making sauce.

4. The executive chef wears black pants.

5. I find this job is attractive.

Exercise 4

Suppose answer:

1. I want to be a coffee shop kitchen sous-chef.

2. We want to be commis chefs.
3. You want to be an Italian kitchen sous-chef.
4. He wants to be a demi chef.
5.They want to be chef de parties.
6. She wants to be a pastry & bakery sous-chef.

Unit 2 Kitchen Facilities

Lead in（导入）

Activity 1 P1. oven P2. microwave P3. gas stove P4. grill

Activity 2 1.T 2.T 3.T 4. F

Focus on learning（聚焦学习）

Part One

Activity 2 P1. salamander P2. deep-fryer P3. dish washer P4. steamer

Activity 3 Q1. It's a deep-fryer.
Q2. It is used for frying food.

Activity 4 B A C D

Part Two

Activity 2 P1. liquidizer P2. slicer P3. stir machine P4. kneading machine
P5. tenderizer P6. mincer P7. waffle iron P8. bone saw

Activity 3 C A B D

Part Three

Activity 2 P1. ice maker P2. freezer P3. ice cream machine P4. refrigerator

Activity 3 1. T 2. F 3. T 4. T

Activity 4 1. s 2. e,r 3. c 4. e,a 5. o,e
6. i 7. i 8. r 9. k 10. i

Give it off（知识拓展）

Practice 1 1.B 2. D 3. C 4. A 5. E

Practice 2 1.What do you call this in English?

It's a deep fryer.

2.What is it used for?

It's used for frying food.

3.How shall we do with the oranges?

You can liquidize them.

4.Refrigerator is a machine to keep food.

Unit exercises（单元练习）

Exercise 1 1. C 2. D 3. D 4. C 5. C

Exercise 2 1. The gas stove is useful in the kitchen.

2. Ice maker is a machine to make ice.

3. Tenderizer is used for tenderizing tough meat.

4. We can use microwave to heat food.

Exercise 3 炉灶设备 gas stove deep-fryer grill steamer oven salamander microwave

机械设备 slicer waffle iron mincer liquidizer stir machine tenderizer bone saw kneading machine

制冷设备 refrigerator freezer ice maker ice cream machine

Exercise 4 1.Liquidizer can liquidize vegetables and fruits.

2.We can tenderize beef with the tenderizer.

3. 这个用英语怎么说？明火焗炉。

4. 我们要用炸炉来做什么？

5. 搅拌机是用来做什么的？

Learning tips

B √ D √

Unit 3 Tools and Utensils

Lead in（导入）

Activity 1 P1. frying pan P2. braising pan P3. sauce pan P4. stockpot

Activity 2 1.T 2.F 3.T 4. F

Focus on learning（聚焦学习）

Part One

Activity 2 P1. sauté pan P2. grill pan P3. crepe pan P4. sugar pan

Activity 3 Q1. You can fry meat, fish, eggs, etc.

Q2. Cook the chicken in butter. Then fix the sauce to go with the chicken.

Activity 4 B D A C

Part Two

Activity 2 P1. mallet P2. whisk P3. spider P4. fish poacher

P5. mixing bowls P6. soup ladle P7. roasting trays P8. grater

Activity 3 C D B A

Part Three

Activity 2 P1. cutting board P2. pot rack P3. kettle

P4. cover P5. roasting fork P6. frying basket

Activity 3 1.T 2.T 3.F 4.T

Activity 4 1. e,a 2. e,r 3.t 4.m,m 5.i

6.a 7.o 8. c,k 9. c 10. a

Give it off（知识拓展）

Practice 1 1. C 2. E 3. A 4. B 5. D

Practice 2 1.What shall I do first?

Cook the chicken in butter.

2.Where shall I put the hens?

Put the hens in the stockpot.

3.What is the next?

Put some water into the stockpot.

4.Shall I beat these eggs?

Yes, with a whisk.

5.What can I do with the frying pan?

You can fry meat with it.

Unit exercises（单元练习）

Exercise 1 1.D 2.B 3. D 4. D 5.C 6.A 7.C 8.A 9.D 10.C

Exercise 2 1. The frying pan is used to fry meat, fish, etc.

2. I need a cutting board to cut meat.

3. I am going to flatten the meat with a mallet.

4. I want to make omelette in the frying pan.

5. He is going to melt butter in the sauce pan.

Exercise 3 1.K 2. C 3. G 4. B 5. A 6. L

7. H 8. D 9. E 10. I 11. F 12. J

Exercise 4 1. Sift the flour with a sieve.

2. Brush the egg yolk on the cake with the brush.

3. 用撇沫器清洁油的表面。

4. 我将如何处理这些薯条？把它们放入油炸篮里。

5. 我将如何处理这些胡萝卜？沥干它们。

Learning tips

C A B D

Unit 4 Kitchen Knives

Lead in（导入）

Activity 1 P1. spatula P2. oyster knife P3. paring knife P4. chopping knife

Activity 2 1. Spatula 2. Chopping knife 3. oyster knife 4. paring knife

Focus on learning（聚焦学习）

Part One

Activity 2 P1.chef's knife P2. cheese knife P3. boning knife P4. salmon knife

Activity 3 Q1. No. The chef's knife is for many different things.

Q2. Mike is going to cut up some meat and cheese.

Activity 4 C B D A

Part Two

Activity 2 P1.whetstone P2.sharpening steel P3.knife sharpener

P4.electric knife sharpener

Activity 3 B C D A

Part Three

Activity 2 P1.carving knife P2. fish scissors P3. pizza cutter

P4. bread knife P5. roasting fork P6. peeler

Activity 3 1. u,l 2. e,e 3. o 4. e,i,l 5. a,l,o

6. p,a,u,a 7. e,c,r,i 8. a 9. a,r 10.y,e,r

Give it off（知识拓展）

Practice 1 1.C 2. E 3.D 4. B 5.A

Practice 2 1. You may use your cheese knife to cut up cheese.

2. I need to/want to chop up some bones.

3. What shall I use to sharpen the knife?

You may use a whetstone.

4. What can I do with it?

You can sharpen knives.

Unit exercises（单元练习）

Exercise 1 1.T 2. F 3. T 4. F 5. F

Exercise 2 1. A paring knife is used to peel fruits.

2. Chef's knife can cut up many things.

3. The electric bone saw is much safer.

4. The oyster knife looks very funny.

5. The chef is cutting some vegetables.

Exercise 3 P1. We cut pizza with a pizza cutter.
P2. We cut bread with a bread knife.
P3. We cut cheese with a cheese knife.
P4. We slice salmon with a salmon knife.
P5. We lift the meat with a roasting fork.
P6. We cut fish with a pair of fish scissors.
P7. We open oysters with an oyster knife.
P8. We lift cake with a spatula.

Exercise 4 1. Please chop the carrots into small pieces with your chef's knife.
2. —What shall I use to peel the potatoes?
—You may use the peeler. Here you are.
3. 剔骨刀是用来剔出肉里的骨头。
4. 你可以用厨刀切很多东西，因此它们非常实用。
5. 请用鱼剪剪开鱼肚。

Exercise 5 课本上出现的刀具

paring knife 水果刀	oyster knife 牡蛎刀
chef's knife 厨刀	salmon knife 三文鱼刀
carving knife 雕刻刀	pizza knife 比萨切刀
chopping knife 砍刀	cheese knife 奶酪刀
boning knife 剔骨刀	bread knife 面包刀

Learning tips

sharp, Cut, Store, carefully, downwards
knife, leave, falling, opener, pocket

Unit 5 Condiments and Spices

Lead in（导入）

Activity 1 P1. sugar P2. salt P3. chill oil P4. honey P5. ketchup P6. vinegar
Activity 2 1. d 2.e 3.b 4.a 5.c

Focus on learning（聚焦学习）

Part One

Activity 2 P1. salad oil P2. butter P3. olive oil P4. mustard
P5. jam P6.soy sauce P7.chicken powder P8. pepper powder

Activity 3 Q1. Yes, it is.
Q2. It can be put in some bitter beverage.

Activity 4 A C D B

Part Two

Activity 2 P1. cinnamon P2. clove P3. dill P4. thyme
P5. bay leaf P6. rosemary P7. saffron P8. fennel

Activity 3 D A C B

Part Three

Activity 2 P1.nutmeg P2.sage P3.basil P4.mint P5.marjoram P6.tarragon

Activity 3 1. T 2. F 3. F 4. T

Activity 4 1. a 2. a 3. a 4. a 5. a
6. e,e 7. t, t 8. p 9. l,l 10. e

Give it off（知识拓展）

Practice 1 1.E 2. D 3. C 4. B 5. A

Practice 2 1.What is the flavor of sugar?
It is sweet.
2.What is condiment?
It is a substance that gives food strong flavor.
3.They are five types of condiments.
4.What is it called?
It is called bay leaf.
5.What is it used for?
It's used for seasoning.

Practice 3 1.This is vinegar.
It's sour.

2.That is salt.
It's salty.
3.They're bay leaves.
They're bitter.
4.We call it chili oil.
It's spicy or hot.

Unit exercises（单元练习）

Exercise 1 1. D 2. A 3. B 4. C 5. A

Exercise 2 nutmeg is not only nutty, but also warm and slightly sweet
fennel is often used in egg, fish and other dishes
sage is often used in cooking meat and vegetable soup
saffron contributes a yellow-orange coloring to food

Exercise 3 balsamic vinegar 意大利香脂醋 champagne vinegar 香槟醋
tarragon vinegar 他力根香醋 grape vinegar 葡萄酒醋
apple vinegar 苹果醋 white vinegar 白醋

Exercise 4 1.There are five types of condiments.
2.What is the flavor of it?
3.Sugar is very sweet.
4. Sage is often used in cooking meat and vegetable soup.
5.What shall I call it in English?

Exercise 5
1. 糖的味道是什么？
2. 香叶略带苦味，辛辣强烈。
3. 调味品使食物味道浓郁。
4.They're five types of condiments.
5.Sage is often used in cooking meat and vegetable soup.

Unit 6 Fruits and Nuts

Lead in（导入）

Activity 1 P1. honey melon P2. watermelon P3. pear

	P4. cherry	P5. peach	P6. strawberry
Activity 2	1.cherry: red	2.peach: pink and green	
	3.watermelon: green and black		
	4.strawberry: red	5. pear:yellow	6.honey melon: yellow

Focus on learning（聚焦学习）

Part One

Activity 2 P1. grapefruit P2. Chinese date P3. lemon P4. plum
P5. grape P6. blueberry P7. apricot P8. pomegranate

Activity 3 Q 1: We'll make grape juice today.
Q2: Put them in the liquidizer.

Activity 4 B A D C

Part Two

Activity 2 P1. pineapple P2. mango P3. lychee P4. kiwi fruit
P5. coconut P6. dragon fruit P7. papaya P8. star fruit

Activity 3 C A B D

Part three

Activity 2 P1. walnut P2. hazelnut P3. peanut
P4. cashew P5. almond P6. chestnut

Activity 3 chestnut, walnut, peanut, hazelnut

Activity 4 1.a,e 2.e,a 3.e,a 4. a 5.e,e
6. i,a 7. a,e 8.y,e 9. a,a,a 10. a,e

Give it off（知识拓展）

Practice 1 1.B 2. A 3. C 4. E 5. D

Practice 2 Fruits:
grape, blueberry, cherry, strawberry
pineapple, dragon fruit, peach
mango, kiwi fruit
Nuts:
peanut, hazelnut, chestnut

walnut, almond, cashew

Practice 3 1. washing the peaches./I'll make peach juice.
2. purple/Yes, I do.
3. Yes, I do./Usually in summer.
4.papaya/ It's sweet.

Unit exercises（单元练习）

Exercise 1 First, wash the fruits and peel the kiwi fruits and oranges.
Next, remove the bottoms of the strawberries and seeds of the honey melon.
Then, cut them up and put them in a mixing bowl.
After that, put ice-cream in it.
Finally, mix them up.

Exercise 2 1.make 2. soak 3. peel 4. remove
5. liquidize 6. put 7. dice 8. slice

Exercise 3 1. D 2. A 3. B 4. C
5. D 6. D 7. B

Exercise 4 1. apple 2. banana 3. grape 4. cherry

Unit 7 Vegetables

Lead in

Activity 1 P1. Chinese cabbage P2. onion P3. celery P4. carrot P5. lettuce

Activity 2 1. carrot 2. Chinese cabbage 3. lettuce 4. onion

Focus on learning（聚焦学习）

Part One

Activity 2 P1. cucumber P2. pumpkin P3. lotus root P4. spinach
P5. eggplant P6. sweet potato P7. parsley P8. taro

Activity 3 Q1. Give him some carrots and onions.
Q2. In the mixing bowl.

Activity 4 D C A B

Part Two

Activity 2 P1. pea P2. red bean P3. soybean P4. chickpea
P5. green bean P6. cowpea P7. haricot P8. broad bean

Activity 3 D C B A

Part Three

Activity 2 P1. artichoke P2. olive P3. turnip P4. red pepper P5. broccoli
P6. asparagus P7.zucchini P8. mushroom P9. yam P10. cauliflower

Activity 3 1.e,o 2.u,i 3.e,u 4.o,i 5.a,a 6.u,o,o 7.u,u,e 8.i
9.i,e 10.e,e 11.a,o 12.u,o 13.a,e 14.e,a 15.a

Give it off（知识拓展）

Practice 1 1.D 2.A 3.E 4. C 5.B

Practice 2 1.Shall I put them on the chopping board?
Well, you'd better put them in the mixing bowl.
2.Shall I cut onions and carrots?
Yes, slice them finely.
3.What will you do?
I'll dice the potato.
4.Please put some salt on the broccoli.

Practice 4 1. Pour 2. add 3. put...on 4. grind 5. slice

Unit exercises（单元练习）

Exercise 1 remove the seeds from the pumpkin
dice the cucumbers
chop the parsley finely
grind the garlic

peel the zucchini

slice the onions finely

scrub the sweet potatoes carefully

cut the red peppers into two

trim off the ends

split the carrots down the middle

Exercise 2 1. Lettuce is the "heart" of most of salads.

2. Are you slicing pumpkins now?

3. Shall I dice the yams?

4. Please chop the onions for the salad.

5. How shall I cook the artichoke?

Exercise 3 cucumbers, celery, onions, chopping board,

carrots, broccoli, mixing bowl, parsley, vegetable

Exercise 4 1. We need some finely chopped onions for the soup.

2. Shall I dice the celery now?

3. He is grinding the garlic.

4. First, wash the carrots, and then slice them please.

5.What shall I do with the artichokes?

Exercise 5

根茎类蔬菜：sweet potato, yam, taro, turnip, carrot, pumpkin, lotus root, spring onion, zucchini, asparagus

叶类蔬菜：lettuce, parsley, celery, Chinese cabbage, spinach

豆类蔬菜：pea, soybean, red bean, green bean, broad bean, cowpea, chickpea

Unit 8 Meat

Lead in（导入）

Activity 1 P1. beef P2.pork P3.chicken P4.mutton

Activity 2 1.pork 2.mutton 3.chicken 4. beef

Focus on learning（聚焦学习）

Part One

Activity 2 P1. sirloin P2.tenderloin P3.T-bone steak P4.rib eye steak

Activity 3 Q1: Sirloin.

Q2: Yes. It’s the most tender part of the beef.

We usually sauté or grill it.

Activity 4 A D B C

Part Two

Activity 2 P1.pheasant P2. deer P3. pigeon P4.goose P5. hare

P6. turkey P7. boar P8. guinea fowl P9. quail P10. duck

Activity 3 C A B D

Part Three

Activity 2 P1.foie gras P2.Osso Bucco P3.oxtail P4. veal kidney

Activity 3 1.T 2.F 3.T 4.T

Activity 4 1.e,e 2.u,o 3.i.e 4.u 5.o,o

6.u,r 7.i,o 8.a,i 9.e,e 10.o,r

Give it off（知识拓展）

Practice 1 1.B 2. A 3. C 4. E 5.D

Practice 2 1.What shall we do with the tenderloin?

We usually sauté or grill it.

2.How about the rib eye steak or T-bone?

It’s more delicious to roast or grill it.

3.Let’s taste typical Beijing food.

4.Is it the most tender part?

Yes, that’s right./No, I don’t think so.

Practice 3 Group1: tenderloin, sauté or grill.

Group 2:chicken, simmer or braise.

Group 3:oxtail, make oxtail soup.

Group 4: “waiji”（外脊）, make it into steak.

Practice 4 1.fry 煎 2. roast 烤 3 . stew 烩 4. grill 扒 5. braise 焖 6. sauté 炒

Unit exercises（单元练习）

Exercise 1 1. C 2.C 3. B 4.A 5.B

Exercise 2 1.What, do 2.more delicious, the most delicious 3.Why not
4. It’s not necessary to, I’ll 5. I’d like to, to 6. Let’s try

Exercise 3 1. C 2.D 3. B 4.A 5.D 6.A

Exercise 4 1.I like beef best.
2.The duck is a little tough. It’s better to braise or simmer it.
3.Why not try to make beef steak by ourselves?
4. 没有必要把肉烤这么长时间。
5. 我想请你品尝一下法式鹅肝酱。

Learning tips（学习提示）

1.E 2.A 3.B 4.G 5. F 6. H 7.C 8.D

Unit 9 Seafood

Lead in（导入）

Activity 1 P1. white bait P2. salmon P3. trout P4. mandarin fish P5. sturgeon

Activity 2 1.T 2.F 3.F 4. T 5.F

Focus on learning（聚集学习）

Part One

Activity 2 P1.caviar P2. sardine P3. grouper P4. flat fish
P5.tuna P6. cod fish P7. anchovy P8. perch

Activity 3 Q1. We can see fish forceps, fish scaler and fish scissors.
Q2. You can scale the fish with the scaler, gut the fish with the scissors

and take out the fish bones with the fish forceps.

Activity 4 B A C D

Part Two

Activity 2 P1. scallop P2. mussel P3. crab P4. lobster
P5. prawn P6. oyster P7. conch P8. abalone

Activity 3 A C D B

Part Three

Activity 2 P1. bream P2. mackerel P3. eel P4. herring

Activity 3 1. F 2. T 3. F 4. F

Activity 4 1.s 2.s 3.s 4.a,a 5.a 6.a 7.o 8.i 9.a 10.e

Give it off（知识拓展）

Practice 1 1. E 2. B 3. D 4. C 5. A

Practice 2 1.What did you eat for dinner last night?
I ate salmon.
2.What shall I do with these fish scissors?
You can get the stomach of the fish with them.
3.What shellfish would you like to eat?
I'd like to eat crab.
4.Could you tell me how to treat shrimp?
Let's de-head and de-vein.

Unit exercises（单元练习）

Exercise 1 1. cut open the stomach of the fish
2. scale the fish
3. bone the fish
4. cut the fish
5.open the oyster

Exercise 2 1.D 2.D 3.D 4.C 5.B

Exercise 3 1.C 2.B 3.D 4.E 5.A

Exercise 4 1. Scale the fish with the fish scaler.

2. Please cut open the stomach of the fish with the special scissors.

3. I'm making the oyster soup.

Exercise 5 Ingredients: onion, flour, mussel, water

Condiments: salt, butter, pepper powder, curry powder

Process:butter, onion, Add flour, mix, mussles, water, salt, pepper,minutes